THE POND BOOK

Valerie Porter

CHRISTOPHER HELM

London

© 1988 Valerie Porter
Line illustrations by David Henderson
Christopher Helm (Publishers) Ltd, Imperial House,
21–25 North Street, Bromley, Kent BR1 1SD

ISBN 0–7470–2607–6

A CIP catalogue record for this book is available from the British
Library

Printed and bound by Billing and Sons Ltd, Worcester

CONTENTS

LIST OF FIGURES

Chapter 1
SKYWATER:
THE EARTH'S EYE

A field of water betrays the spirit that is in
the air. It is continually receiving new life
and motion from above. It is intermediate in
its nature between land and sky.
Henry David Thoreau: **Walden**

Do you remember your first glimpse of the sea? Do you remember the magic of its tantalising gleam in the distance catching your eye through a gap in the scenery? Water is light; it is irresistibly attractive.

The sea inspires with its vigour, its promise of danger, its unknowable depths and expanses; but an inland lake, by the fact of its finitude, is less daunting and can soothe the spirit on a quiet day as it calmly reflects blue and grey skies. Those who watch lakes can find solace in solitude.

A pond is an intimate, personal lake. The child splashing through a rain-puddle, or plucking a stranded worm from its muddy edge, soon learns that a pond is a bigger, more promising puddle, full of textures and living things, a reservoir of sensations and curiosities. Do you remember tadpoles and minnows, whirlygig beetles and water boatmen? There is so much *life* in a pond.

Nobody has actually counted the nation's ponds but every local or regional survey reveals that they are disappearing faster than hedgerows. Most of our ponds originally had a purpose; today, unyielding pipes and taps and tanks have displaced them and, as is the nature of ponds, they are quietly and rapidly being reclaimed by the land, neglected, literally swamped in silt and growth, until in many places all that remains of their presence on earth is an isolated fringe of willows, a slightly damp patch in a field, or the persistent annual visitation of toads to a housing estate.

And as the ponds shrink and tiptoe to their deaths, so too does the abundance of life they supported. Your children today might ask, 'What is a steam train?' Your grandchildren might one day ask, 'What is a pond?'

What is a Pond?

A pond is a small freshwater lake. There is no rule which dictates the size at which a body of water becomes a lake rather than a pond but, broadly speaking, lakes are large and deep enough to provide a stable habitat whereas ponds are shallow enough to allow the penetration of light to most of their depths. This means that they warm up or freeze over quickly, and can lose water quite noticeably in hot weather or fill fast when there is a lengthy downpour. Thus the life that colonises a pond must be tolerant of a fluctuating environment or be able to adapt by mobility.

Sunlight is the great generator of life so that a pond, well penetrated by its rays, is potentially a much richer world than a lake and supports a far greater diversity and density of living organisms per unit volume of water and unit area of bed. But it is a fragile world, easily disrupted by nature and even more vulnerable to man. However, man the destroyer also has the ability and option to preserve, conserve and create — and that is what this book is all about.

The Origins of Ponds

One of the fascinations of ponds is their great diversity. They reflect not only the nature of their surroundings but also the will of man, because in very many cases ponds are artificially created, either for a definite purpose or incidentally as a result of some other human activity. Yet most ponds (except those which are still managed for specific productive reasons) ignore their origins and welcome local wildlife without discriminating about their own natural or unnatural pasts. Pond life is largely determined by local circumstances: the type of soil and underlying rock affects the quality of the water and the nature of the pond's silt as well as the terrestrial vegetation that grows in the region which, in turn, affects the water. The climate of the area and the topography play their part, too.

The great majority of natural ponds and lakes in Britain are in the northern and western regions, especially the uplands. The first requirement of a body of still water is a depression in which it can lie, whether it be a pot-hole in a farm track or a fiord-like chasm carved out of landlocked mountain valleys. The most powerful architects and excavators of British lakes were the massive glaciers that creaked their way across the Ice Age landscape, colossal diggers of frozen water grinding at weakened rock, ploughing through faults, scraping great heaps of mountain debris ahead of them and dumping haphazard dams of boulders and scree. Thus, on the older, harder rocklands of Britain, frozen water was the instrument that dictated where in the future the melted water would run or would linger in lochs, tarns and cwms. Water in these regions tends to be on the scale of lakes more than ponds, though the hills and moors have many acid, soft-water pools.

Other ponds had less awesome origins. Many are bywaters — a section of once-flowing water cut off from its own stream, perhaps by a sudden landslide or by the leisurely process of meandering and silting that creates oxbow lakes and rivermouth lagoons. Less often, a pond basin is created when a pocket of soft substance subsides or is dissolved — perhaps a salt dome collapses or a weak patch of soluble limestone is eroded.

Another natural pondmaker is that industrious engineer, the beaver. Although now drastically reduced in numbers and restricted to North America and parts of Europe and the USSR (in all of which it had to be reintroduced after its virtual eradication), the beaver was originally widespread in the forested temperate regions of the northern hemisphere, including Britain. In manipulating the environment to suit itself, by building and maintaining dams across watercourses, it created lakes and ponds that became home for many other creatures as well. Its effect on the landscape was dramatic and considerable.

Man, however, has had far more effect than beavers have. Although some ponds are shrivelled natural lakes which have gradually silted up with organic and inorganic debris, the majority of ponds (in lowland Britain, at least) are man-made and reflect the nation's history for those who have the urge to trace it in the landscape. A pond's original purpose can be unearthed by a mixture of local research, common sense and imagination, and the first step is to look at an Ordnance Survey map to see if a particular pond or swampy area could form part of a larger system, or is near the site of some long-forgotten building or industry. Clear away the accumulated knowledge of today: ignore existing woodland, buildings, roads and boundaries, and plunge back into an uncluttered past. Imagine who might have wanted a water supply on the site and what they might have required of it.

In the days when man was insignificant, the natural ponds of Britain were used by the wild animals. Families of aurochs (the ancestor of domesticated cattle) came to the water's edge to drink where the banks of lakes, bywaters, rivers and streams were gentle enough to give them access. Their trampling feet, aided by those of the wild boar of the forests, eroded the banks and gradually created broad, shallow pools. The daintier hooves of deer and sheep helped to consolidate the sloping margins, and the drinking animals were as easy game for the native wolves as are today's vast African herds for the big cats. Man, too, came to drink, and each wave of immigrants naturally followed the watercourses or settled by springs and pools, letting their domesticated livestock make use of the wildlife drinking places where they could.

As man became more in control of his environment, he began to dig wells and also to put water where it was needed, rather than relying on its natural distribution and often seasonal availability. Water is a resource which can be managed — stored, directed, circulated. As well as exploiting the natural waters man began to create, first by manipulation and, much later, by methods so remote from the natural source that many people today half believe that water can only come out of taps. They have little notion of its sources, or of its journey from source to tap. Like wildlife and the weather, water becomes remote from day-to-day living.

There was a ubiquitous dampness to lowland areas; much of the land was marshy and the ground was waterlogged for at least part of the year. 'Wetlands' were extensive; broadly speaking, where lowlands were not wooded they were wetlands, though often enough the forests were also squelchy underfoot. People cleared the forests for farming and then began to tackle the problems of draining the potentially productive fens and marshes. Land drainage on any scale was not technically feasible until about the seventeenth century, and in many parts of England large areas remained undrained until the nineteenth or even twentieth century. With that drainage, the landscape changed and become more uniform — more *organised*: there were straight lines (hedges, ditches, water channels) with neat edges, rather than relaxed natural curves and variety. Rivers became 'systems'; their meanderings were ironed out, their periodic floodings (so beneficial to meadowland) were regulated, their margins and shallows were eradicated. Natural ponds and man-made watering places were drained, filled in, ploughed over, and replaced with pipe systems and tanks which echoed the severe straight lines of humans that are so alien in nature and so rigid. Among man's straight-line orderliness, there is little room for other life.

But that is today. Yesterday such extremes had not been reached.

The most obvious use of a pond is as a supply of drinking water for humans and other animals and the great majority of ponds did serve this purpose, whether they were natural reservoirs gradually broadened and 'puddled' by the trampling feet of wildlife and livestock, or were created by deliberate damming or digging. In many cases a spring or stream feeds drinking ponds and quite often those who decide to turn a damp patch in the corner of a field into a wildlife or landscape pond find that, buried under the silt and plantlife of centuries, there is an unsuspected spring struggling to yield its clear water.

Some of the old drinking ponds lie beside lanes and greenways, and they were used by drove herds on their way to market or pasture, and by carters whose horses needed watering. Some of these ponds today seem to be in the middle of nowhere but often give a clue that what is now an overgrown, impassable track was once a more important route to somewhere. More often, drinking ponds are near an old farmyard or out in the fields; some were dug primarily to help drain the fields and were then incidentally used to water the animals.

A few of the pasture ponds are shaped vaguely in the form of a cross or clover-leaf. Occasionally these are duck decoy ponds: wildfowl were attracted to the open water at the centre and were lured down an arm of the pond into netted funnels from which they were unable to escape. More often, these star-shaped ponds were livestock ponds, probably dug when the huge old open fields became enclosed and the movements of the animals were restricted. The ponds were sited where several pastures met: each arm of the pond served a different field and the centre was deep enough to act as a 'wet fence' that kept each group of animals in their respective areas. This was probably effective with sheep, who are none too keen on paddling, let alone swimming, but cattle are natural wallowers and the water at the centre of the pond would have to be quite deep to deter them.

Water as primarily a barrier is, of course, the essence of the moat and many large manors and farmhouses were moated for protection from enemies or to keep livestock in (or out!). Today, those moats may linger on as unusually shaped ponds, outlining a steading which has long since vanished or has been absorbed by the village that grew up around the manor. Castles, as every child knows, also had moats and drawbridges and so did some of the monastic establishments.

The monasteries were generally large-scale landowners and farmers, and as such had considerable influence on the shaping of the landscape. Many monasteries were built near water and the monks were famous not only for their wool production but also for farming fish in quite elaborate systems. They built special 'stewponds' (the word is derived from the French *etui*, meaning a case or sheath, based on the Old French word *estui* which sounds as if it has connections with 'estuary', which in turn comes from a Latin word meaning 'boiling, commotion' — which takes you back to 'stew' and conjures up quite a vision of seething fish!). The famous Serpentine in London's Hyde Park was originally a terraced series of monastic fishponds built in the thirteenth century, with the waters of the Westbourne river flowing through ponds arranged in descending steps. At its deepest, the Serpentine is about 9m (30ft) though most of the water is no more than about 5.5m (18ft) deep, and its bottom is clay. It is a long lake now, rather than several ponds, and there are still fish in it.

On the borders of Hampshire and Sussex, where a tongue of Wealden clay is edged with a horseshoe of Lower Greensand, there is a pond marked on contemporary and older maps as 'Fishpond'. It is in the grounds of an imposing old house which was originally the site of a small monastery and later of a romantic rural retreat built by King Charles II for one of his mistresses. It is said that he used to swim in the pond as a boy, and indeed he had a lifelong fascination for ponds: his collection of ornamental wildfowl on the lake of London's St James's Park was the historical basis of the huge collection there today. The rural Sussex fishpond, however, was on a much smaller scale and, today, it supplies trout to a hotel which has succeeded the romantic retreat. The pond is the largest and lowest of a chain of pools controlled by a weir and fed by three springs emanating from a wooded, sandy, south-facing hanger which provide very pure drinking water for the hotel's guests. The headstream runs beneath the building to emerge in a series of ornamental falls and garden pools before the water enters the old fishponds which are but the first of a series of ponds that continue downstream: a millpond, a village pond, an old sheep dip, duckponds, cattle ponds, a hammer pond and so on down the broadening valley to join other streams, each with their own ponds — watercress beds, flight ponds, estate lakes, old ice-making ponds near secret ice-houses, and angling lakes as well — all entering an ancient marshland drained into submission in the nineteenth century but with a defiant water table still lingering only a little way below the top layer of the valley's clay and remembered in the names that show the place as this Marsh and that Marsh. Derelict pumphouses hide in the woods; they are silent now but once their regular beat was heard ceaselessly as they pumped water up the valley to the big houses and estates on the hangers.

This is my valley, my home, and it gives a microcosm of ponds within the three or four miles between the springs and the river.

The millpond half a mile downstream from the fishpond's splash is typical of its kind and a good example of how and why ponds disappear. The mill, the site of which was recorded in the Domesday Book, was a working cornmill until the early part of this century and the tall, narrow stone building is now a private house, the old wheel displaced and forlornly rusting. The beautiful gardens make the most of the bowl of land in which the house sits, with the level of the impounded millpond at first-floor level, its waterhead held back by a large, solid clay dam of some antiquity only a stride from the back windows. The millpond is now half its original size, the upper end having gradually silted up over the last 50 years, first populated with reeds, then gradually invaded by trees (planted cricketbat willows and spontaneous alders and oaks) until now a third of the original pond is decidedly terrestrial, its earth firm underfoot and thriving daffodils blooming in their masses in spring. There is still a marshy area between daffodils and clear pond but even here the shrubs, pampas grass, balsams and brambles are slowly ousting the bulrushes and increasing the land area at the expense of the old pond. Long ago, someone took out the felled elm trunks that held the banks in place, and thus hastened the development of swamp and marsh as the banks crumbled. The old millrace has been channelled into a series of decorative waterfalls and the sluice that controls the overflow is gradually decaying, its iron plates rusted so thin that they threaten to spring out of place in times of spate.

The remaining area of the pond is half its original depth and the deep silts are home to a population of sizeable swan mussels and innumerable invertebrates. Streaks of roach and rudd swim above them, but no trout have found their way downstream from the fishpond. A gabble of mallards dabbles the edges, weakening the dam at its angle with the bottom. Patches of waterlilies brighten the water and, in spring, bright green globules of algae, like glistening jelly marbles, rise from the silt and block the outlets.

Right up to the end of the eighteenth century, waterpower was an essential source of energy and mills driven by water served many purposes other than grinding corn into flour. The recorded history of watermills goes back to at least the first century BC, in ancient Greece, and the technology was brought to Britain by the Romans. The Domesday Book in 1086 recorded mills as valuable assets of great local importance, and in many cases the same sites still have at least a millpond, if not a mill, even today.

Although corn-grinding was their prime function in most ages and regions, wheels turned by water could also supply power for all kinds of industries, perhaps the earliest of which were the twin Wealden glassworks and ironworks, situated where there was waterpower to drive the machinery and an endless supply of wood for furnace charcoal, as well as the raw materials of iron and silica in the ground. A mile downstream from the cornmill there is an old hammer pond, five times larger than the millpond, which would have built up a considerable head of water to drive the powerful, heavy, nodding tilt hammers that forged the iron and the bellows that stoked the furnaces.

Today, loitering secretly in the heart of the woods, Hammer Pond is quiet — strangely quiet, indeed, and not just for lack of industrial noise: its broad waters have suffered in the last two decades from agricultural run-offs further upstream. These accumulate where the water is held back by the massive stone horseshoe weir before it thunders down 6m (20ft) into a bottomless round pool, where once the hammers would have sung but now there is only the crash of water, soothing in its monotony, and the kingfisher darting as brightly as any blacksmith's sparks. Above the pool a tiny cottage, once the home of ironworkers, is now a retreat for nuns who minister to the needs of a nearby Buddhist monastery, and the setting is ideal for the meditations of monks.

Waterwheels in other parts of the country drove woolmills, papermills, snuffmills, gunpowder mills, coppermills, furniture factories — the applications were endless. Indeed, the presence of water for power led to the development of whole villages where before there had only been empty valleys. Water gave impetus to industry, and industry needed people — lots of them. In fact nearly all villages, before and after the Industrial Revolution, tended to be sited near a good water supply — a stream, river or good spring and, in due course, a pond, which lies at the heart of any Englishman's dream village.

Between Hammer Pond and the millpond in this Sussex valley are various 'rural' ponds scattered around and in the village. Several livestock ponds are really no more than widenings of the stream at certain points, with gently sloped banks giving the animals access to water, but the old village pond is more purposeful. The 'village' here did not in fact exist until quite recently; the area was far too wet, though there were many hamlets scattered around on islands of drier land or on the slopes of the sandy hangers where two or three large estates built cottages for their staff. There were several farms and, as a point of reference, the valley was bisected by the north/south march of a Roman road running between Chichester and Silchester. The 'village pond' lies as close to that road as do the millpond and fishpond; the watercourse runs parallel to the road and about 275m (300yd) from it.

This village pond was originally a sheep dip and it even has a 'plug'. But this does not belittle it as a village pond: most of them were created not just as drinking ponds but also as fire-fighting reservoirs (thatch and wattle blaze well) and 'dips' for sheepwashing and for cartwheels which were soaked in the pond to swell and fit their rims — not to mention 'dips' for those unfortunates who were deemed to be witches or miscreants.

A pond frequently preceded its village and was part of the reason for the settlement developing where it did. Very often tracks developed because natural ponds existed: the track went from pond to pond. In some cases, some say, the Romans made ponds beside their new roads in order to drain off surface water; in others, ponds were dug out beside a track to provide water for travelling horses and livestock, and where two roads crossed was an obvious place for a pond to be dug. The crossroads became an informal meeting place and those with an eye for trade began to set up services there, probably starting with refreshments more substantial than pond water. An inn would be built by the pond, then perhaps a wheelwright and a blacksmith would look after passing carters, and gradually the pond and the crossroads became the centre of a small village. Thus the pond was both the focal point of the eventual community and also the village's *raison d'être*.

Our village pond, like so many others, became neglected during the latter half of this century. In this case it was reverting to marsh, but in many villages a more unpleasant problem is the rubbish which is inevitably dumped into the pond as soon as it ceases to serve a purpose. In the early 1970s there was a nationwide campaign to 'Save the Village Pond', which caught the imagination of many villagers. Later in the same decade, the Queen's Silver Jubilee gave the campaign fresh impetus, and that is when our own reed-choked patch was tackled by a group of local volunteers, young and old, and turned into a Jubilee Pond. Today it is a popular duckpond where mothers, children and old folk meet, chat and gain much simple pleasure from feeding the mixture of wild mallards and crossbred farm ducks which regard the pond as their base. A pair of Canada geese nest every year on one of the small islands, secure from the attentions of local dogs and foxes, rearing their brood of yellow, wide-legged goslings which vaguely try and terrorise the hesitant resident moorhens for a week or two, until the parents march the whole family across the fields to spend a month or two on the millpond.

Canadas, mallards, tufted duck, mandarins, carolina ducks and other waterfowl are commonly seen in this valley, commuting between the various ponds (they are spoilt for choice) and paying special attention to the 'Duck Hotel', a personal pond lovingly created from a waterlogged field whose owner builds dozens of nesting sites and shelters for the wild ducks, simply because she loves the birds. Mallards nest in every cranny of her walled garden, and upstream they nest in the gardens of the millhouse too, at ground level, or in the steep banks above the splash, or in some style among the honeysuckle's tangle on top of the disused millwheel.

Duckponds can be for the pleasure of the birds' company but the older ponds had a more practical purpose. Farmyard ducks and geese were for the table — either their eggs or their meat — and many farm ponds served to accommodate the table birds as well as supply drinking water for livestock, or provide a reservoir for farm irrigation and fire-fighting.

Ducks have also been exploited for sport, especially since the development of firearms, and gamekeepers in recent times have dug out special flight ponds for wild species of duck which are reared on small ponds, often in or near woodland, where they are fed and cared for with great diligence, and then on the day of the shoot they are caught up and transported to the flight pond for release. Naturally, they will fly back to their home pond, and the guns can line up between the two ponds sure of some fairly easy shooting. Unfortunately, a tame mallard which habitually visits a garden pond looks no different in flight to a keeper-reared or genuinely wild duck. . .

Quite apart from this infinite variety of man-made ponds originally managed for a purpose (and no mention has yet been made of more productive purposes — angling lakes, fishfarms, watercress beds and the like), man's various activities have carved incidental ponds and lakes out of the landscape. Wherever materials have been extracted — gravel, clay, stone, slate, sand or peat, for example — the sunken pit or quarry is often left to its own devices and, if its substrate is suitable, soon gathers rainwater and springwater. The scale and siting of the excavation determine the type of wildlife which these unintentional ponds can support, though if they are turned over to 'leisure' (boating perhaps, or swimming, or waterskiing) the wildlife is deterred by human disturbance.

In the Wealden oak forests there are often isolated groups of shallow pits where the clay was dug out for ironworking, pottery, field marl, wall-daub or brick-making, and these secret pools of stagnant water, blackened by decaying leaves from the surrounding trees that exclude most of the light, offer a habitat very different to the wide, sunny expanses of a gravel pit in open country where anglers, speedboaters and herons find their sport. Incidentally, many a house's cellar originated because clay was dug to make the bricks that built the house and in some cases it made sense to build the house over the claypit. Clay, of course, makes an excellent waterproof lining for a pond. . .!

Peat-digging on a large scale led to the creation of those magnificent overgrown ponds, the Norfolk Broads, and also the marshes of the Somerset Levels. The Broads are now under great pressure: sewage phosphates and agricultural nitrates so enrich the water that algae threaten to obliterate all other plantlife and, as the marginal plants cede victory, the banks begin to disintegrate (their demise hastened by the wash of motorboats) and their mud adds to the silt, so that the water is in retreat.

Many of the incidental ponds, unlike most of the purpose-built ones, are not connected with a watercourse and there is one type of pond, carefully created for the specific purpose of watering livestock, which has no convenient stream, spring or watertable to feed it: the dewpond. Dewponds, perhaps because of their name, their circularity and their unexpected siting high up on the downlands, have an almost mystical aura and a cornucopian reputation for always being full, as if by magic, in spite of the lack of a spring. They are found in areas rich in ancient history and folk-lore but in fact many of them were made comparatively recently, that is to say in the last 400 years or so. Nor do they fill with 'dew' (which has so many connotations of fairies, early morning innocence, gossamer and so on). Their source of water is rainfall, either direct from the skies or as it runs off the surrounding land, and in some parts they are called sky ponds, or cloud ponds, or mist ponds. A few do have much older origins — perhaps Anglo-Saxon or even Roman — but unless they have been maintained to prevent the puddled-clay bottom being cracked by drought or by the feet of cattle, they tend to choke up over the years and gradually become nothing more than a slight dip in the turf. That is why so many are of much more recent origin: the ancient ones have vanished.

There *is* some magic to a dewpond but only the magic of knowledge and skill. The ponds were designed, sited and built by craftsmen whose art was handed from father to son and who were careful to protect their reputation by not disseminating the art. Possession of a trade secret gives a man a certain status among his fellows.

Chapter 2
UNDERSTANDING PONDS

Fish say, they have their stream and pond;
But is there anything beyond?
Rupert Brooke: **Heaven**

Ponds can be appreciated simply for the pleasure they bring as living landscape features, or they can be made 'useful' and cultivated for their produce, be it fish, fowl, plants or the less tangible harvests of wildlife or leisure. To gain a deeper understanding of ponds (and they can offer so very much of interest) it is important to know a lot more about that most versatile of the four Greek elements: water, in which life itself originated and in which there is far greater diversity of life than on the land. Water is also a very part of life: the bodies of all living animals and plants contain a substantial proportion of water, typically considerably more than half the total weight of the body.

It is not the intention to give an endless recitation of scientific terms in this book, but a few of the more common ones have to be included and there is a glossary at the end of the book to explain them. Ecology (or 'Ecology': it is sometimes a self-conscious discipline, like 'Conservation') is a vital branch of science but somehow it has become overburdened with labels and is becoming more concerned with terminology and statistics than with *life*.

Ponds are full of life. The world of the pond is essentially dynamic and three-dimensional (aquatic organisms are not tied by gravity to the solid earth) and its energy and elements circulate in a complex web of interconnected cycles through some identifiable zones and layers of habitats. The character of the pond is constantly changing, and the main influencing factors are precisely those four Greek elements: earth and air, water and fire. Yes, fire — that is, the distant, omnipotent source of all energy on Earth: the sun. The sun plays a prominent and vital part in all the cycles within a living pond.

Water reflects not only light but also the nature of substances with which it is or has been in contact and, because of the essential fluidity of water, a pond can be affected by circumstances which may be far removed from its actual site. It depends on where the water comes from and what it picks up on its way to the pond.

The source of all pondwater is the world's oceans. Seawater is evaporated at its surface and rises to form clouds, which in due course shed rain back on the earth. Depending on the geology of an area, some of the rainwater soaks into the ground on which it falls and some runs downhill off the surface until it reaches, or creates, a watercourse or a body of still water like a lake or pond. Watercourses eventually empty into the sea and the cycle continues.

The water which has seeped into the ground (known as *groundwater*) will find its way quite easily through any porous rock or through cracks and channels in harder rock until it reaches the local *watertable*, which is the level beneath which the underlying rock is fully saturated and at which water will stand in a well. Typical porous media are sands, gravels and sandstones, through which surface water will drain quickly and which tend to make the water a little acid (because of their silicon content). Chalk is an 'aquifer' (a 'water-carrying' or permeable rock) and water also travels along fissures in limestone or in hard rocks like granite. Rainwater falling directly onto impermeable surfaces will of course run off the surface and travel overland, absorbing soluble substances on the way. In some conditions, where surface run-off water has nowhere to go, the area becomes waterlogged.

The level of a local watertable varies. For example, it rises after prolonged rainfall, or sinks lower in times of drought or varies because of human interference like artificial drainage somewhere in the catchment area or water extraction on a large or widespread scale. Ponds which depend on the watertable level are at the mercy of such agents and are more than likely to dry up from time to time.

If the groundwater seeps down to water-retarding layers like clay, or solid impermeable rockbeds without any fissures, it will flow along the top of the layer until it reaches a point where the strata are cut by a slope, and the water issues as a *spring*.

If water becomes trapped *under* an impermeable layer, pressure gradually increases as more water accumulates and the setting is ripe for an *artesian well*, i.e. a borehole through which water rises under its own pressure, without having to be pumped. The London basin is a good example: rainwater falling on the exposed chalk of the North Downs and Chilterns permeates the chalk and continuously accumulates in the basin under overlying clays, ready to be tapped when a borehole is sunk through the clay to the trapped watertable.

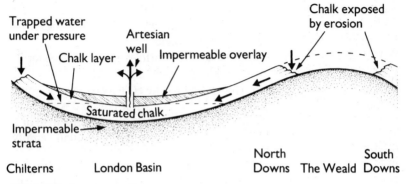

Figure 2.1 *Artesian well*

Rain falling on insoluble, hard rocks feeds the upland streams and the water is as pure and soft as the rain itself. No natural water is absolutely pure, however, if purity means containing nothing but the basic elements of water (hydrogen and oxygen). Even rainwater contains other dissolved gases like carbon dioxide and nitrogen, and increasingly it also includes industrial pollutants like sulphur dioxide, ammonia and perhaps nuclear fall-out dust. Sulphur dioxide, emitted by the burning of fossil fuels both in the home and in industry, greatly increases the acidity of rain and at certain levels 'acid rain' can have an adverse effect on life, including life in ponds and watercourses. But stillwaters can also become acidified in the presence of pyrites (mineral or iron sulphides), for example in newly excavated pits for extracting coal, mineral ores, sand or gravel: pyrites exposed to the air are oxidised by microbes, producing sulphuric acid, and the water held in such pit-lakes is like dilute acid, in which plants and animals cannot flourish.

Acidity and pH Values

The degree of acidity of water is a crucial factor in determining its ability to support life, because it directly affects the chemical reactions that are so integral to biological and metabolic processes. Each different type of living organism has a specific range of acidity in which it can survive and, in many cases, the range is very narrow so that water of greater acidity is not only intolerable to the species but often lethal.

The acidity of any solution is measured in terms of its *pH value*, which is the concentration of dissolved hydrogen in the solution.

Technically, a pH ('potential hydrogen') reading is the reciprocal logarithm of hydrogen-ion activity but no mathematical skill is needed to find out whether or not water is acid. 'Neutral' water has a pH value of 7.0, acid water less than 7.0 and 'basic' water more than 7.0. A 'base' is a substance that reacts with an acid to form a salt and thus neutralises the acid.

The pH scale ranges from 0 to 14 but it is unusual to find pondwater as acid as 3 or 4 and anything less than a pH of 4.5 would be considered very acid. In general, fish can tolerate a pH range of about 5 to 9.5 but different species have much narrower ranges in which they thrive and in most cases the best reading will be about 7.0 or 7.5. Above all, fish need a stable pH, and basic waters tend to be more stable than acid ones.

If the pH of water is below about 5.7, many invertebrates find conditions intolerable either because it is directly lethal to them or because the food on which they feed is sensitive to low pH and is therefore absent — especially some forms of algae. In addition, acidity inhibits the decomposition of organic debris on which many invertebrates also feed.

Another problem with acid water is that it releases metals held in the soil and some of these, like aluminium, could be harmful to aquatic life. Substances like lime and chalk, which are basic, can neutralise the acid and thus harmful metals remain locked in the soil.

Acid water tends to taste a little sour or salty and is often found in upland tarns on hard rock substrates like granite.

Alkalinity

Acidity can be neutralised by increasing the concentration of the water's *alkalinity*: the greater the alkalinity, the greater the capacity to neutralise acids, which means that 'basic' water is in fact alkaline. Alkalinity is more or less equivalent to the concentration of bicarbonate ions in the water, as opposed to carbon dioxide: if there are more carbonates the water will be alkaline, but if there is an excess of carbon dioxide it will be acid. Carbonates and bicarbonates mix with acid in water, diluting the acidity; carbon dioxide, on the other hand, dissolves in water to form carbonic acid, and the pH of water can be influenced by carbon dioxide from the air or from the respiration of aquatic lifeforms. Thus the pH value can fluctuate. Calcium carbonate and bicarbonate are perhaps the most common factors increasing alkalinity in water; in fact calcium is the dominant ion in determining pH. It is also an essential nutrient for plants: alkaline waters are much richer in life than acid ones. Plants and algae thrive, and therefore so do the creatures that depend on them for oxygen and food. The most productive waters have pH values between 7.5 and 8.5, and an alkalinity of 50 to 220 ppm is the most productive for fish.

The Acid Test

There are some very simple methods of actually measuring the pH of water, usually by means of chemical reactions causing colour changes, or more sophisticated methods involving the measurement of tiny electrical charges. Most people have memories of litmus paper in the school laboratory — those little strips which turn red in acid solutions

Table 2.1: Plants in acid and alkaline waters

The followng plants may be found in or by particularly acid or alkaline waters, either by preference or because they are able to tolerate such conditions:

Very Acid (pH<4.5)	Acid (ph<7)	Alkaline
Bulbous rush	Bogbean	Hornwort
Floating bur-reed	Bogmoss	Reedgrass
Quillwort	Cotton grass	Shining pondweed
	Floating clubrush	Water celery
	Jointed rush	Water chestnut
	Lesser spearwort	Watercress
	Red pondweed	Watermint
	Shoreweed	Water soldier
	Various-leaved pondweed	
	Water lobelia	
	Water milfoil	

or blue in alkaline ones. More sensitive are the indicator papers or colorimetric solutions which give gradations of colours indicating the degree of acidity, when compared with a guide chart. Simple soil-test kits bought in garden centres work on the same principle. For those with a more serious interest in the quality of pondwater or feeder streams, it is advisable to use an electronic pH meter (which is based on the effect of hydrogen ions on a thin glass membrane) or have laboratory tests carried out by, say, the local water authority or environmental health officer, who can also test for other factors and for the presence of various trace elements.

Plants and animals found in the water, or trees and shrubs on surrounding land, can give an indication as to whether the water or soil tend to be acid or alkaline.

Hardness

Alkaline waters are often 'hard', and acid ones 'soft', though not necessarily so. In general, however, low alkalinity = low pH = low hardness = soft acid water.

Hardness is due to the concentration of soluble salts in the water — mostly of calcium or magnesium, which are both important elements for the majority of life forms. Again, calcium carbonate is probably the most important factor in water hardness (it is also known as carbonate of lime); it is insoluble unless carbon dioxide is present, and then it changes into soluble calcium bicarbonate. Thus hard water is also alkaline.

Hard water often leaves traces of lime at the waterline when the pond level drops, and of course hard water deters soap from making a good lather. Soft water can be hardened by the addition of lime.

If rain falls on a soluble rock like chalk or limestone (which are calcium carbonates), or on soils, it will of course absorb dissolved minerals and any springs from such strata will contain those minerals. The physical hardness or softness (in texture) of the underlying rock determines to a large extent the chemical hardness or softness of the water but perversely (to the layman) hard rocks tend to yield soft water, and soft rocks hard water. The softest waters usually run off the hardest rocks, like granite, and soft water is also found on slates and sandstones. Basalt water is on the hard side; chalk a little harder and limestone more so. Very hard water is usually influenced by marls (limey clays), gypsums and dolomites.

The effects of calcium carbonates on water hardness are:

Soft   <80mg calcium carbonate per litre of water
Medium hard   100–200mg/litre
Hard   300mg/litre

Oxygen

A crucial factor in the acidity of water is its carbonic acid content. Carbonic acid is dissolved carbon dioxide, one of the two dominant gases affecting waterlife. The other is oxygen. The two gases are exchanged during the course of respiration and the balance between them, particularly in the enclosed environment of a pond, is critical. Aquatic plants must have access to inorganic carbon in the water in order to synthesise carbohydrates, and their main sources of carbon are dissolved carbon dioxide and carbonate ions. This manufacturing process is photosynthesis and it crucially affects oxygen levels in the pond. During active photosynthesis, aquatic plants absorb carbon dioxide from the water and release oxygen, which is life to all other pondlife as very little oxygen is absorbed from the atmosphere at the surface of the water. However, photosynthesis only takes place in the presence of light, building up maximum levels of dissolved oxygen during the day, but at night (or where plants have no access to light) it ceases and no fresh oxygen is released. But plants continue to respire in darkness, as do the pond's animals. Respiration is a process during which oxygen is absorbed and carbon dioxide is released and therefore the available oxygen dissolved in the water is rapidly depleted and carbon dioxide levels are increased. In extreme cases this can lead to the impairment or death of organisms in the water and any scheme of good pond management must seek to maintain adequate oxygen.

Oxygen levels can be diminished by, for example, high temperatures. The warmer the water, the less its capacity for carrying dissolved oxygen (DO). For example, at 0°C (32°F) it will hold 14.6mg DO per litre of water; at 10°C (50°F) 11.3mg; at 20°C (68°F) 9.1mg; but at 30°C (86°F) only 7.6mg — almost half the rate at 0°C (32°F), and a level which will kill trout and not do much good for salmon.

The balance of living matter in the pond is important for oxygen levels, too. Animals use up oxygen in respiration and give out carbon dioxide, so if there are not enough plants (especially if the plantlife is suddenly decreased) there will be inadequate oxygen for the animals. If there are no plants, there is no photosynthesis, therefore no oxygen release and thus no aquatic animal life. On the other hand, a sudden increase in the plant population (for example, algae blooms) can release huge quantities of oxygen all at once, but because these explosions tend to result in dense mats of vegetation at or near the surface, light is cut off from plants below so that they cannot actively photosynthesise and must use oxygen from the water, thus depleting the oxygen supply. The surface plant layer, rather than replenishing the water's dissolved oxygen, loses most of its released oxygen into the atmosphere and the net result is a sudden and possibly fatal drop in oxygen levels in the lower waters.

The situation can be saved by aeration techniques which stir in oxygen from the air or circulation systems which pump fresh oxygenated water through the pond. Fishfarmers sometimes resort to beating the water with rods, or paddling about in a rowing boat, or churning the surface with an outboard motor.

Oxygen levels vary enormously over any 24-hour period. For example, they may drop in cloudy, close weather because the poor light reduces photosynthesis and the warmth of the water reduces its capacity to hold oxygen. Another dangerous situation is when good sunshine has encouraged a real burst of floating plant growth, blocking light for submerged species, and by about four o'clock in the morning critically low ebbs are often reached.

The best way of judging oxygen levels in the water is by watching how pondlife is reacting (yes, fishfarmers do sometimes go out in the early hours of a summer morning to check the four o'clock syndrome). A sudden change of water colour from greenish to clear is one warning sign that planktonic plants are failing to produce adequate oxygen and the fish, at first lethargic, may soon be gulping for air at the surface of the water. No fish can tolerate less than 4mg of oxygen per litre of water.

Temperature

Light — sunlight – is thus crucial to the living pond. The sun also influences water temperature, of course. Temperature fluctuations are less extreme than in air (how else could Muscovites and Serpentiners go swimming in midwinter?) but in a small, shallow body of water they can be severe enough to be a major influence on the life of a pond. The shallowness usually means that there is not much difference in temperature between marginal waters and the depths, so that there is nowhere for creatures to escape to. In deeper lakes, there can be quite a substantial difference in temperature in different parts and layers of the lake, but ponds are rarely much deeper than the height of a tall man or two and the different light and temperature zones (and therefore life-niches) tend to be dictated by organisms in or near the water. For example, light penetration can be inhibited not only by algae blooms but also by surface mats formed by the tiny floating leaves of duckweed, or by shade from overhanging trees, or by suspended inorganic matter, or a richness of organic life in the water which renders it turbid. A degree of turbidity is a sign of a productive pond but if the water is too clouded the light cannot reach the lower waters and, again, the oxygen balance will be upset. In extreme cases, where the bottom becomes anaerobic (devoid of oxygen), there will be little life except a prolific crop of tubifex worms and other organisms able to survive with little or no oxygen. Any oxygen in the mud is quickly used by bacteria to break down decaying matter on the bottom.

Most animals, especially fish, have fairly specific temperature ranges in which they can thrive and beyond which they may not survive at all. The first signs of distress in overwarm conditions will be reduced feeding activity and general lethargy, and the problem is complicated by the accompanying decrease in oxygen levels — the indirect effect of heat.

Ice

Water reaches its maximum density at about 4°C (39°F). As its temperature drops below that level, it becomes lighter. Thus ice, which is formed at 0°C (32°F), is light and floats to the surface: the density falls to 0.917 at that point. Ice is a poor conductor of heat, so that any unfrozen water beneath the floating ice will retain its heat more easily, and plants and creatures will be able to survive, unfrozen, under the layer of ice. Only a very shallow pond will freeze right through.

But there are problems other than being frozen. For example, if a snowfall covers the surface of the ice it will prevent the penetration of light into the unfrozen water, so that the plants will be unable to continue photosynthesis. Oxygen levels will decrease; carbon dioxide will increase; and there will be no exchange of gases to the atmosphere through the ice barrier, so that there can be a build-up of noxious gases under the ice. Always sweep snow off a pond, especially if there are fish in the water.

Colour

The colour of water gives clues about its quality and about the life it supports. Ignore the reflected blue of the sky, but pure water does have its own clear, transparent blueness, especially when it is devoid of life. In water, purity is not life-sustaining! Aquatic plants cannot live on hydrogen and oxygen alone, and without plants there can be no other organisms: no tiny animals, no deadmatter converters like bacteria and fungi, and the food and energy cycles are broken before they even begin to form.

An almost Mediterranean blueness can be seen in very clean chalk ponds, and a less blinding blue in mountain lakes.

Where there *is* life, there is also death and decay. Decaying plant material (humus) tinges pure water with clear shades of brown, yellow and green. In winter, a lowland pond will often have a clear greenish look to it, which is a healthy sign of life. (Remember to observe the colour of the water: don't be misled by the colouring of the pond bottom.)

If the bottom is deficient in oxygen, anaerobic bacterial activity on decaying matter produces gases, like methane and hydrogen sulphide, which not only smell but also turn the mud blackish. Many a shallow woodland pool is little more than a dark grey gasworks where fallen leaves accumulate and slowly rot.

If fertilising substances like nitrogen and phosphates have been artificially introduced in excess, the water is likely to look dull and turgid, with an unpleasant darkish slime on the pond's muddy bottom indicating oxygen deficiency.

Living organisms also colour water, especially the algae. Blue-green algae, when they 'bloom' in warm sunshine, can turn the whole pond green or red. Others give it a yellow look. Sometimes a population explosion of certain crustaceans can colour the water, too. A particular type of bacteria which oxidises iron compounds creates a bright rust-coloured area, especially where an emergent spring rises from iron-rich ground.

Pollution

Ponds are sitting targets for abuse. There they are, confined worlds of water, just ripe for invasion by undesirable agents from the air, over the surface of the land, or along the water sources that may feed them — side streams, overflowing watercourses, springs, unsuspected seepage of underground water, and so on. And a pond has no defence against these invasions, unless a helping hand is willing to follow an alert eye. Flowing water, in rivers and streams, is also susceptible but at least the very flow helps to disperse unwanted substances and thus dilute their effect over the whole length of the watercourse, or until a barrier is met. Many ponds do not have the advantage of self-cleansing currents.

'Pollution' is one of the catchwords of our age and, although water pollution has always occurred, it is only in the last century or two that the activities of the human race have increased the likelihood of pollution to such a huge degree that disasters can now occur in hours rather than be the result of many years of accumulating trouble. It only takes a thoughtless moment for the entire life of a pond to be wiped out, eradicated, just like that, as if someone had exploded a nuclear bomb in its heart.

Air pollution of one kind which can affect a pond has already been mentioned — sulphur dioxide pumped into the atmosphere by the burning of fossil fuels and then precipitated in rainwater into the waiting pond. The water pollutants that normally come to mind are chemicals, poisons, detergents and suchlike — toxic substances emanating from industry which are theoretically under control, though accidents still happen and the effects of past abuse are still accumulating. Oil and petrol spillages, either direct into the water or as run-offs into watercourses from roads, or waste oils poured deliberately into drainage systems, quite simply smother aquatic life by spreading a film over the water surface. In quantity, the effects are disastrous and lethal. The smaller the pond, of course, the more exaggerated the effect and a typical village pond next to the road can be devastated by road run-off, especially when the road is given fresh tarmac.

Salt used on winter roads can kill freshwater fish and other life that cannot tolerate salinity if it runs off into a pond. Obviously, substances used for killing weeds, insects, fungi and other 'pests' are bound to affect aquatic life if they find their way into the water, which they all too easily do.

Far more dangerous, partly because their effect is not perhaps recognised by many people, is pollution from agriculture and from sewage. The major agricultural pollutants (apart from pesticide sprays which have a direct killing power) are silage effluent above all, and also manure and the run-off of fertilisers and lime spread on fields.

The unwitting leakage of a small amount of silage liquid trickling through cracks in the pit and finding its way into the ditches and thence to streams and rivers can murder fish in their hundreds. A bigger leakage is catastrophic and dramatic: water inspectors talk of a 'sea of silver' or a 'silver river' and those who have seen it will never forget the sight of every fish in the watercourse downstream from the leakage lying belly up on the surface, a great sheet of dead fish covering the water and floating on the current to pile up in their thousands against the first barrier. The effects in a static, enclosed environment like a pond should be compulsory viewing for every livestock farmer: what may seem like a minor nuisance of maintaining a properly sealed silage pit or slurry pit could save the life of an enormous area of water. If downstream activities include any kind of fish husbandry, the guilty livestock farmer might be lucky to escape a lynching and certainly deserves to be on the wrong end of a substantial insurance claim.

Silage effluent is 15 times more polluting than cow manure and 150 times more polluting than untreated domestic sewage. A 400T silage clamp can produce as much pollution as the sewage system from a town of 150,000 people and a ton of fresh silage can produce up to 455 litres (100 gallons) of effluent.

The word 'pollution' carries strong overtones of uncleanliness, contamination and foulness, all of which suggest harm. Quite often, however, the initial effect of a water pollutant is to encourage life and stimulate growth, and this is the case with silage and sewage. The problem is that the stimulation is excessive: there is an explosion of life, certain organisms run riot and the whole delicate balance of the aquatic environment is destroyed. A common example is when an input of nutrients (say, from fertilisers) gives a pond's plant population a veritable feast and there is a substantial increase in vegetation. That may seem desirable, but excessive growth quickly leads to overcrowding and the greater the density of plant life, the greater the demand on the pond's overnight oxygen supply; and this demand is compounded, because plant growth on or near the surface can so blanket the pond that sunlight is denied to any but the upper layer of plants and, consequently, photosynthesis becomes limited or impossible beneath the mat of green and the oxygen balance is further depleted. When the plant matter dies and decays, there is far too much debris for the confined world of the pond: huge amounts of oxygen will be used up as the bacteria work to break down the sudden depositing of dead material.

A continuous level of enrichment, even at quite low levels, can lead to too vigorous plantlife which gradually encourages the pond to silt up and eventually become nothing more than marshland. Fertilisers, perhaps spread on adjacent farmland to increase crop productivity, can too easily be washed into watercourses or farm ponds and there should always be a buffer zone of unfertilised land between field and water. Such fertilisers typically include nitrogen and phosphates, both of which promote aquatic growth as much as that of corn. All kinds of plants, including algae, flourish on it. Indeed lack of phosphates in a natural environment is the main inhibitor of aquatic growth, so that a balance is normally maintained.

Table 2.2: Pollution

Obvious signs of pollution (whether eutrophic or toxic) include dead fish. oily slicks, foam, smelly gases, vigorous duckweed and algae, invasive bulrushes, a change in water colour or turbidity, or a change in the species which inhabit the water. The more polluted water becomes, the fewer different species there will be, and those that do continue to thrive are often seen in very large numbers. In conditions which are ideal for, say, a conservation pond, the water will support a wide variety of species and none of them will be overwhelmingly dominant. In extreme pollution there will be no life of any kind, but otherwise look for an *abundance* of certain species which will give some indication of the degree or lack of organic pollution.

Heavily polluted

'Sewage fungus' (spongy growth of colonies of various bacteria,
   algae and microscopic animals)
Dense blanketweed and algae blooms
Fennel pondweed
Sludge worm
Rat-tailed maggot

Species which will tolerate some pollution

Arrowhead	Blackfly larvae	Carp
Bulrush	Bloodworm	Perch
Broad-leaved	Caddisfly larvae	Roach
pondweed	Great pond snail	
Curled pondweed	Ramshorn snail	
Duckweeds	Leech	
Reedmace	Water boatman	
Yellow waterlily	Water louse	
Water starwort	Water slater	

Intolerant species (little or no pollution)

Alternate water	Mayfly nymph
milfoil	Stonefly nymph
Quillwort	Flatworm
White waterlily	Freshwater shrimp
Watercress	Mussels
Water soldier	Trout
	Water shrew

If there are plenty of the following species, the water is
unpolluted (or only slightly polluted) and will also supply
plenty of food for fish and waterfowl:
   Mayfly nymphs, dragonfly and damselfly nymphs,
   caddis larvae, freshwater shrimps, water slaters, water
   hog lice, water beetles and their larvae, water bugs,
   water boatmen, water mites, leeches, snails, flatworms.

Pond Ecology

In its course from the source to its final dispersal into seas, lakes or marshes, a stream changes character considerably. At first it tends to be clean and fast-flowing, with lots of oxygen in the water but very little plantlife because the newly sprung water lacks nutrients, and because the speed of the water current denies a foothold or resting place for plants except where they can cling to the rocks. In due course, the stream begins to widen and to accumulate organic debris, minerals and silt from the land so that its average rate of flow decreases and the plantlife can begin to colonise its waters. The plants in turn hinder the current, so that the further downstream the more silt, vegetation and therefore dependent animal life there is. This impediment of flow is one of the major effects of exuberant plant growth in a watercourse and can lead to a raising of the stream level and of the watertable in surrounding areas, with a risk of local flooding which allows the watercourse to be recharged with nutrients from the land as the floodwater drains back into the stream.

The speed and regularity of waterflow vary in different parts of any cross-section of a stream. Where the water runs by the banks or along the bed, its flow is naturally inhibited by friction between the water and the surfaces that confine and channel it, whereas in clear midstream the water can run on more freely and therefore faster. Where the flow is reduced by friction, deposits can begin to accumulate and plants can find a home. They anchor themselves with their roots. The types of aquatic plant that float freely will be washed downstream with no chance to linger unless they find a bywater — perhaps a pond.

Many ponds are fed by a stream or spring so that they, too, have a current, however slight, and thus they often behave like an expanded section of stream, with quieter margins where plants take root and, again, as the vegetation increases the root systems trap more silt and debris and in due course their own dying leaves add to the debris. Thus a habitat develops for more vegetation until, eventually, the pond itself is reduced to a wet patch under all those plants. Left to their fate, and given conditions in which plants can thrive (adequate nutrients and light, and no catastrophic pollution), any pond will naturally regress in this way, and disappear. All ponds long to be land.

There are several major plant groups in a pond, suited to different habitat zones, as Figure 2.2 shows. Note that ponds offer an extra dimension: plants can rise from a soil surface (the bed) like any

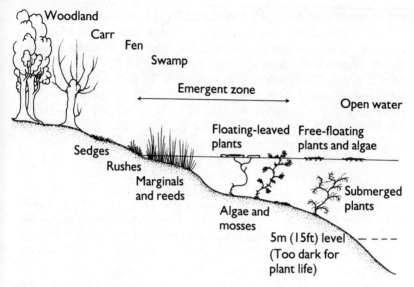

Figure 2.2 *Plant zones*

terrestrial plant or can dangle their roots in the water with no contact with the land at all, floating freely on the surface of the pond and grabbing the best of the light. But it is the plants whose leaves and stems are totally or partially submerged which are of vital importance to other pondlife, because it is these which oxygenate the water at all levels, especially the fully submerged and truly aquatic species, including many algae.

Plants produce plenty of decayed matter which increases the silt layer and, in due course, the bed slowly rises. The shelving by the banks accumulates enough silt to give purchase to reeds and rushes which can tolerate waterlogged soil; and the more gentle the original bank slope, the greater the opportunity for invasion by these terrestrial plants. Reeds are to wetland what birches are to dry: they are supremely successful invaders and great opportunists, and the reed army soon encroaches on the pond's margins, advancing its hold because the reed roots trap more silt and bind themselves into an impenetrable mat, building up the shelf bit by bit. The pond is in retreat. On the landward side of the reeds, the soil is less swampy; it becomes fenland, suitable for sedges, and beyond them the willows and alders begin to colonise the damp soil. Each group gradually moves towards the disappearing pond, taking advantage of the drier conditions produced by the preceding group, until, finally, only the carrs of willows and alder are left as tombstones for a pond that quite literally sank without trace.

Such a situation is the tendency for lakes and ponds which are *eutrophic*, a word which means 'well fed', that is, the water contains plenty of nutrients, whether from natural sources or as a result of human activity in agriculture and industry. In fact, when applied to lakes and ponds the term eutrophic means 'overfed'. Nearly all lowland ponds are (or become) eutrophic and the condition is generally self-increasing because the nutrients increase plant growth, which means that there is an increased volume of decaying organic matter in due course, which releases more nutrients during the process of decay and so on, the pond becoming increasingly rich and crowded like the stomach of someone who has eaten too much. The land adds its share to the meal: substances are washed out of the soil or run off the surface and enter feeder streams or go directly into the pond itself. Agricultural fertilisers and livestock manures all contribute their goodness in run-offs and there is also the occasional accidental leakage of highly concentrated nutrients in silage effluent or sewage, which are so stimulating that they ultimately kill or at least damage much of the life in a pond — a severe case of indigestion.

In a naturally eutrophic pond the increase in vegetation is more gradual; algal blooms make regular appearances in response to warm, sunny weather, but are not so overwhelming that other life has to struggle to survive. Increasingly, human activities are hastening eutrophication.

Eutrophy equates with high productivity and is generally the final stage in a pond's life before it regresses to swamp and eventually becomes more or less dry land. There are several less productive stages and a pond may run the gamut of them during its existence or may remain fairly constant. The most barren waters are described as *dystrophic*, and these are unproductive or of only very low productivity. *Oligotrophic* waters are of low productivity, and medium levels are achieved by *mesotrophic* ones.

Thus the natural succession (progression or regression, depending on the point of view) is for a clean, empty, dystrophic pool to proceed through the trophic categories to eutrophism, and thence to the wetlands of reed swamp, fenland and carr, and finally to dry land.

Cycles and Chains

Unless plant life can become established, a pond will not support any animal life. Aquatic plants need some basic 'ingredients': essential nutrients (like calcium, magnesium and phosphates), inorganic carbon, dissolved oxygen and the energy of light. The major elements required are carbon, hydrogen and oxygen, which the plant can convert by photosynthesis into carbohydrates. The latter can be used to make protein if the plant can take up nitrogen and sulphur, which need to be present in various nitrates and sulphates.

Among living organisms, only plants can actually create carbohydrates from the raw materials; animals cannot and must eat plants (directly, or by eating those who do eat plants) to obtain nutrition. The basis of all food chains, therefore, is the plants' manufacturing process, photosynthesis.

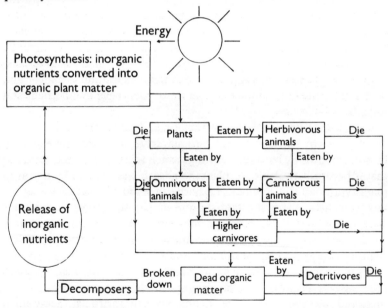

Figure 2.3 *Food chains in a pond*

The sun's energy, therefore, gives impetus to that most basic of the pondlife cycles, the food chain. Microscopic plantlife is eaten by microscopic animal life and the two form the pond's *plankton*. Larger animals feed on the plankton and on larger plantlife (macrophytes); some of them are in turn devoured by carnivorous creatures, and so on. All these organisms, whether plant or animal, eventually die and their organic remains, along with the manure of living creatures, are consumed by detritivores and broken down by decomposers, including fungi and bacteria, which release nutrients from the decaying matter (using up some of the pond's oxygen in the process) and the cycle continues, with the sun's energy enabling the plants to synthesise fresh organic matter from the released inorganic nutrients by means of photosynthesis. The pond is almost a world in itself.

There are endless permutations within these food chains. In a typical pond, for example, planktonic plants like algae are grazed by tiny crustaceans like waterfleas and rotifers, and these creatures are preyed upon by certain fish, which may be eaten by larger carnivorous fish, which might be taken by a heron, a mink or an angler, so that their energy and nutrients are removed from the pond. Pondweeds might be eaten by swans or ducks, which also dabble for small crustaceans, snails, insects and worms and might themselves succumb to man or fox, or lose their ducklings to pike or heron, or their eggs to moorhens and rodents.

Dead fish, dead plants, dead crustaceans, dead anything in the water will be eaten by detritus feeders (various larvae and mites, for example) or broken down by bacteria and fungi which release inorganic matter so that it becomes available for plants to absorb and convert by means of photosynthesis. But it should be borne in mind that, at each stage of any food chain, a very substantial proportion of the energy stored in the food species is lost: the predator only gains about ten per cent of it.

All this eating, and dying, and most of all living and breathing, requires energy and oxygen intake, and it is little wonder that oxygen levels in a pond can fluctuate so dramatically. Those who manage ponds must always be aware of the danger and take steps to ensure that dissolved oxygen levels are maintained by artificial aeration if necessary. The aim, as in all aspects of pond management, is to achieve a balance in all respects.

A pond is so vulnerable. Because it is shallow and confined, its waters are subject to sudden changes in temperature, quality, living organisms, levels, and so on, and these changes place considerable stress on the aquatic lifeforms that cannot instantly move to a safer, more stable environment.

Those lifeforms, in a typically eutrophic pond, are of infinite variety and their lives are interlocked in complex webs, each having effects on others in a constantly dynamic environment. Whole books have been written about life in ponds and only general information can be given in this book. Later chapters look more closely at some of the plants and animals that are likely to be found in a pond.

Chapter 3

CREATIVE PLANNING

> *One may not doubt that, somehow, good*
> *Shall come of water and of mud;*
> *And, sure, the reverent eye must see*
> *A purpose in liquidity.*
>
> Rupert Brooke: *Heaven*

Ponds, as the preceding chapter makes clear, have a compulsion to revert to dry land and for that reason they need to be managed if they are to survive as ponds. Management, however, can only succeed where there is a sense of purpose, of knowing not only what you are doing but why you are doing it.

This book is intended to give both practical guidance and inspiration. It may be that the reasons for creating a new pond have already been decided, but for some people there will be a vague desire to 'do something with water' and the flexibility of water gives the imagination plenty of scope. Table 3.1 suggests some of the possibilities about what can be done with water and it might stimulate some new thoughts. There is indeed a 'purpose in liquidity' and quite often such purpose goes unrecognised.

Table 3.1: Possible ponds: some ideas

Conservation/Nature

General pondlife	Farm nature trails
Amphibian ponds	Pond-dipping for children
Dragonfly ponds	Birdwatching wetland area
Rare species reserves	Town 'lungs' and pocket parks
Village ponds	Part of larger nature reserve
Wildfowl ponds	

Pleasure and Leisure

Angling: coarse fishery or trout lake
Ornamental waterfowl
Domestic ducks
Shooting
Garden goldfish pond
Garden nature reserve
Garden ornamental pond
Chinese watergardens
Japanese watergardens
Fountains and cascades

Public park lake
Estate landscaping
Focal point for a
  housing estate
Leisure park — lake
  for boating,
  swimming, water
  skiing, etc.
Waterside pub or
  restaurant

Purposeful Ponds

Aquaculture cycles (livestock manure/algae and weed/fish)
Edible algae and pondweed crops
Growing ornamental waterplants for sale
Watercress beds
Osier willows/rushes/reeds for basketwork, thatching, etc.
Cricketbat willow crop
Trout farm
Other fishfarming: carp, tilapia, mixed table fish
Specialist fishfarming: ornamentals, angling stock, fishfarm fry
Crayfish farming
Eels
Farm ponds: livestock drinking supplies, fire-fighting reservoir
Irrigation (agriculture, glasshouse, orchard)
Domestic water supplies
Wet fence/moat
Waterpower
Dewpond — for livestock drinking water or for conservation

A common realisation, among those who value ponds for whatever reason or purpose, is that water is a resource — an asset — whether for wildlife, for leisure, for visual or spiritual refreshment, or for purely practical purposes like irrigation, supplying drinking water or aquaculture of various kinds.

Many stages in pond-building are similar regardless of the purpose of the new pond, and this chapter describes basic techniques. The requirements of special-purpose ponds, including their design and management, are considered later but in every situation the first priority is to decide why a new pond should be created and to have a very clear idea of the aims. Then the planning can begin, and it is essential to plan carefully and thoroughly, well in advance of any work on site. The liquidity of water gives it the power to affect and be affected by distant events and situations as well as those in the immediate surroundings: what one person does with water is bound to affect other people, and a multitude of other living creatures and plants, and sometimes the very fabric of the landscape. It is a power that needs to be wielded with due consideration and certainly with a sense of responsibility.

For example, if conservation of wildlife is an important factor it should be borne in mind that creating a new pond, or even reclaiming a derelict one, might destroy an existing habitat for, say, a rich community of bog plants. Wetlands can be just as important as areas of open water. In conservation, incidentally, it is generally much better to renovate an existing pond (preferably in gradual stages using gentle techniques) than to build a new one, and this renovation work is considered in more detail in Chapter 5.

Most ponds can be attractive landscape features and serve wildlife as well as being practically or commercially exploited. In most cases straight lines, regular shapes and concrete show a severe lack of imagination and an arrogant disregard for the countryside and its inhabitants, though in some intensive aquaculture systems they may be considered necessary for economic reasons and ease of management. Over-efficient aquaculture, however, smacks of greed except in situations where it serves to feed a community living close to the breadline. Food production, though essential to human survival, does not necessarily have to be at the expense of other living organisms; nor do we live by loaves and fishes alone. We need the aesthetic and emotional pleasure that water can bring.

Site Assessment

Quite often, a pond site suggests itself. There is a damp pocket of land, or a spring, or a stream course or a natural valley which somehow feels right for a pond. However, the sites of some new ponds have to be determined by the purpose they serve, regardless of the natural landscape.

If old maps, names or memories reveal that once upon a time there was indeed a pond on the site, then in all probability it will be suitable for a new one as long as the reason for the disappearance of the original can be determined. It may have been something more serious than disuse and neglect: perhaps the ground itself is polluted in some way, or the water source has vanished or become unreliable. While the maps are out, look for plans of field drainage systems which, if they run through the proposed site, will need to be diverted to prevent damage to the system and to prevent the pond from being drained or polluted.

In an initial assessment, look at the geology and soil type and check indicator species for clues. Spend a lot of time checking the water source because the availability and quality of water are fundamental to the life of any pond.

Substrata

The permeability of the site's underlying rock naturally affects the holding capacity of the pond. Check substrata by making a borehole to take a sample at least 1m (3ft) down. Dig it by hand or use a small mechanical back-acting excavator, which can be used down to perhaps 3–5m (10–15ft). Such boreholes can provide material samples for expert analysis and can also be used to gauge the watertightness of the site. A rock bottom is not necessarily waterproof, even if the rock itself is impermeable: there may be fissures in it and the pond will need a clay lining.

If the site is on impermeable soils like clay, the pond can probably hold water without special lining and its capacity will be adequate as long as the supply of incoming water, from whatever source, equals or exceeds the natural rate of evaporation and loss by seepage, usage or controlled outflow. Evaporation rates depend on the ratio of water surface area to depth: a wide, shallow pond loses water more rapidly than a deep, less expansive one. Seepage rates depend on the soil or the liner.

On permeable soils like sands and gravels, it may still be possible to keep an unlined pond reasonably full if the watertable is high, but in most cases it will be necessary to make the pond bottom impermeable. Dig test pits at several points on the site to check its waterholding capacity. Dig them in pairs and fill one of each pair with water, leaving the other to fill itself. Watch the pits over a week or two: if the filled pit drains quickly, a liner will be needed. The level in the self-filling pit will give an idea of the natural watertable level. If constancy of watertable supply is important, run these trials for a year before starting work on the pond because it may be that under certain conditions the level drops so much that the pond will dry out. A simple method of checking the level of watertable down a test hole is to put a weight on a length of string, rub coloured chalk along the string, lower the weight into the water then take it out and measure the length of string which has retained its chalk coating.

Water Sources

Some ponds can exist without any obvious water supply other than rainfall; others act as catchment pits for run-off from the surrounding area and many are fed by underground seepage. Obviously, such ponds are subject to waterlevel fluctuations and potential pollution from surface dressings like fertilisers, manure and crop sprays. Many are affected by watertable levels, which are subject to rainfall (or its lack) and to drainage or extraction activities. Unsuspected sources of pollution might include domestic or industrial rubbish tips, where substances seep down into the soil and contaminate the groundwater. There may also be pollution by seepage from domestic sewage pits in rural areas, and the installation of flushing toilets in cottages before mains drainage came to rural areas did in fact lead to the death of many a village pond: cesspools all drained into the ditches and thence the overflow waste-water found its way into the pond. Today, however, the scale of this is less likely to lead to disaster unless a number of pits are very close to the pond, but there is an additional problem in that many cesspools now carry household detergents (though their owners should know better). In comparison, incidentally, waste from an upstream fishfarm could be as potent as untreated sewage from a population of several hundred people.

If the local watertable is high, pondmaking can be as basic as a simple 'scrape' which fills to the level of the table and also absorbs sideways seepage from groundwater as well as overground run-off. Scrapes are often made where the substrata are deep ones of permeable material, or where impermeable matter is overlain by permeable strata. Be aware, however, that the watertable may vary so that the pond could dry out, either seasonally, or permanently if there are certain human activities in the region. Any activity in which a large hole is excavated and pumped out — for example, major building works, road construction, laying watermains and gasmains, mining and quarrying, or of course land drainage schemes (new field drains or someone deepening their ditches) — will have a dramatic and possibly permanently depressing effect on the watertable over quite a wide area. Large new areas of concrete or a new road, especially a motorway, can alter the watertable permanently because rain which used to fall on the ground and seep down to the table now falls on an impermeable surface and is channelled into a mains drainage system instead. A large new forestry plantation will also upset the watertable because of drainage works, and will in addition possibly pollute with the nitrate and phosphate fertilisers given to the new trees.

But those are future events beyond the control of the pondmaker. For the present, dig a test pit down to the watertable and observe it over a whole year to see what happens to the levels in that pit. If the watertable is not high, the pond will need lining.

If surface run-off of rainwater forms an important proportion of the water supply, design and site the pond to gain a maximum catchment area. A dewpond is a good example and embankments around any excavation increase the efficiency of a pond's water collection. Make sure, however, that the site is buffered from agricultural run-offs: a pond slap in the middle of an arable field is wide open to pollution from fertilisers and sprays.

For many purposes, however, a running source is preferable. Running water aerates the pond and freshens the supply, to the benefit of most forms of life in the pond. Indeed, for fishponds it is essential.

Main Watercourses

Rivers and streams can be tapped if they are close at hand. However, they are an obvious source of pollution over which the pondmaker has little control, and the rate of flow can be highly irregular according to the season, the weather and upstream abstractions, so the preliminary survey should include an upstream tour to see where problems are likely to arise. Such sources are not ideal for commercial fishfarming, which needs high-quality water and a reliable supply, and the presence of a suitable source (like a spring) is the decisive factor in siting a fishfarm.

There are various regulations controlling the abstraction or impounding of water from rivers and streams (see Red Tape, p. 44, for details) and prospective pondmakers should consult the local water authority in the early stages of planning. Note that other water users may have weirs which on occasion may be closed or fully opened, often with dramatic and immediate effects on downstream ponds, though the user is duty-bound to warn (and preferably consult) downstream users before such an event.

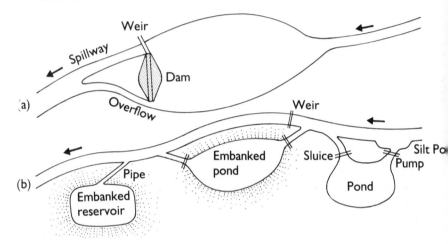

Figure 3.1 *(a) On-stream pond; (b) off-stream ponds*

Ponds depending on watercourses can be built *on-stream*, i.e. the stream is dammed, or impounded, so that its entire flow passes through the pond, or *off-stream*, in which case part of the flow is diverted to the pond but the main watercourse continues on its way unhindered: water is abstracted from the stream and returned to it lower down.

On-stream ponds are at the mercy of the full force of a stream in flood, and it is difficult to isolate the pond from the watercourse should this be necessary (for example, to avoid pollution). They are expensive to build because dams need to be very solid, and they offer less scope and flexibility of design. It is generally better (cheaper, simpler and safer) to excavate a hole rather than to build a dam — that is, to dig down rather than up — so choose a site where the lie of the land allows minimum excavation. Dam-building requires considerable expertise and, if the dam should burst, the results could be disastrous. Sunken ponds don't burst.

True off-stream ponds are independent: they are fed directly by the stream, either by diversion (with sluices controlling the inflow) or perhaps by pumping. The term 'off-stream' is also applied to ponds fed by springs, surface water run-off, direct rainwater, land drainage systems or groundwater seepage on a high watertable. If the pond is big enough, it can act as a reservoir and will withstand a reduced inflow: it should be designed to survive a half-metre (20in) drop in level. Most off-stream ponds are excavated from the centre, with the spoil being used to make embankments that raise the contained waterlevel above the surrounding level, thus increasing the capacity of the pond.

Springs

One advantage of springs is that they often reach the surface under the force of gravity so that there is no need for expensive pumping, as long as the pond is built below the level of the spring. They also generally give a very reliable and pure water supply at a constant temperature, whatever the season. If they are used as water sources for fishfarming, they may lack adequate oxygen and should be allowed to cascade before entering the pond.

Quite often a muddy puddle of a farmpond, once cleaned out, is found to be a spring. Its presence might be betrayed by a lack of plants and animals very close to the actual source, keeping away from the cooler, disturbed water. Some 'springs', however, are artificial land-drainage outlets which can carry surface pollutants and, more alarmingly, seepage from silage or slurry pits.

Wells

If you intend to use or sink a well or borehole, first of all check whether there are other productive wells in the area (as an indication that the substrata are indeed water-bearing). Then bear in mind that extraction of water from wells and boreholes by any party on a large scale, or by many on a smaller scale, could easily drop the levels for all the users.

Water from wells or boreholes will need to be pumped, and this adds to the expense of the whole operation. They can also be expensive to sink in the first place but they do on the whole offer a continuous supply of high-quality water. If a pump is needed it can be powered by hand, by electricity (have an emergency generator in working order in case of power cuts), by diesel, by the wind or by waterpower (see Pumps, Rams and Waterpower, p. 85). It may be worth asking the Institute of Geological Sciences to give advice (for a small fee) on the chances of a new borehole supplying adequate water.

Land Drains

Make use of groundwater in shallow sands and gravels by burying pipes
in trenches below the watertable, following the lie of the land. The job
should be done when groundwater levels are at their lowest — perhaps
September — to ensure that the pipes are laid as deep as possible below
the top of the table. Use unglazed clay tile drains, or porous concrete,
but don't drain from fields which are heavily fertilised. Land-drainage
supplies may vary with watertable fluctuations and can easily become
polluted.

Water Divining

The basic principle of water divining or dowsing (also known as
rhabdomancy — a word which really rolls off the tongue) is to use a tool
as an extension of your own body so that the reactions of the body to the
presence of water are amplified by the tool. The tool is only a tool: the
body itself is the 'receiver', and then it is up to the brain to interpret the
body's reactions.

Whatever method is chosen, the first step is to relax. Let your mind
rest: do not concentrate on the tool but be receptive to it.

Sometimes no tool is necessary, if the body is particularly sensitive,
and some dowsers 'recognise' the presence of water by symptoms like
tingling hands, a sudden stomach pain, cramp, loss of balance,
tightening of muscles, a facial tic or even a yawn. For the less sensitive,
or less practised, tools are needed to exaggerate the body's reactions and
there are several choices and methods of use.

The simplest may be angle rods. Cut two wire coat-hangers as shown
in Figure 3.2 and hold one rod in each hand, upright and loosely in
your fist so that the rod arms can swing and turn in a horizontal plane.
It might be easier to use a cotton reel or piece of wooden dowelling as a
loose handle. Figure 3.2 also shows the neutral and active positions of
the rods.

Alternatively, use spring rods. For these, you need two strips of any
resilient material under tension, so that its springiness gives lack of
stability. The strips, perhaps 20-30cm (8-12in) long, are bound
together in some way to form a Y-shape (or you can use a naturally
Y-shaped twig) and they react in a vertical plane. The classic dowsing
rod is hazel, but it must be fresh and green, up to perhaps 45cm (18in)
long and 3–6mm (⅛–¼in) thick. Hawthorn, cherry and dogwood
could be used (also green) or whalebone, or any similar materials.

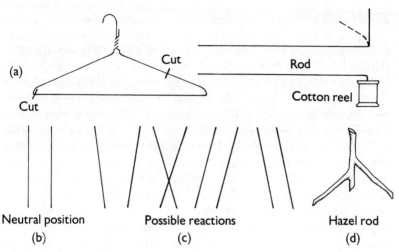

Figure 3.2 **Divining tools**
(a) *Cut two wire coat-hangers to make rods. Use cotton reels as handles in which the rods can rotate freely in a horizontal plane.*
(b) *Neutral position for rods.*
(c) *Possible reactions to presence of water.*
(d) *Hazel rod*

With the palms upwards, hold the spring rods in your fists so that they are under tension. If necessary, increase the tension by twisting the wrists in a horizontal plane. The reaction of the tool to the presence of water will be a vertical movement of the fork of the V.

Another tool is the pendulum, which can be a weighted bob of anything you like on a cord about 30cm (12in) long. Set it to swing lightly on a stable axis back and forth. Its reaction to water will be gyration, and the direction of whatever you are seeking will be indicated by a change in the axis of the swing.

Whether or not divining 'works' has perhaps never been proven scientifically, but certainly some people are better than others at knowing, or guessing, where underground water lies hidden.

Water Assessment

Water sources for practical ponds should be tested for quality, particularly the pH value, hardness, and levels of trace elements or pollution. The water authority will test water samples for a new enterprise but there is a charge for this service (currently about £75 for a very basic analysis, with the price escalating if more detail is required).

Simple pH tests have already been explained (see p.16) and there are other ways of assessing pH values. As an advertisement for a well-known company's pH-meter suggests, there is a particularly sensitive fish, the mormyrid, which in healthy water emits a measurable voltage of electricity in pulses of 400–800 per minute. If water quality deteriorates, the fish becomes uncomfortable and reduces the frequency of its emissions. That is unlikely to be a practical proposition for most pond users, but the point is that pH-meters are based on sensor electrodes: when the electrode is placed in the water, a measurable voltage is set up on its surface and is directly proportional to the pH value of the water, which is then displayed on the meter to an accuracy of perhaps 2 decimal places.

Flow

For some purposes it is necessary to have an idea of the quantity of water entering a pond; indeed, this must be gauged accurately for dammed ponds large enough to require approval by various authorities. The water authorities monitor major river flows on a continuous basis and may give access to their recordings for general interest, but for a specific pond the inflow needs to be measured by means of weirs and the use of standard tables. For an approximate measurement, use the float method; for low flows from a spring try the jug method; but for a more accurate assessment of stream flows you will need a thin-plate weir.

With the *float method*, estimate the average cross-sectional area of the stream (multiply width by depth) and then time how long it takes for a floating object like an orange or apple to move over a measured distance (1–2m, 3–6ft) in mid-channel in calm air conditions. The flow rate is the speed of the object, in metres (ft) per second, multiplied by the area of the stream's cross-section, and multiplied further by a factor of 0.8 to take account of the average difference between surface velocity and the flow speed of water in the channel, where it is reduced by friction. This gives a figure of cubic metres (ft) per second which is the volume of water discharged in a second. It can be converted to litres by multiplying by 1,000 (multiply the volume in cubic feet by 6.25 to convert to gallons). You need to take an average of a series of readings, but there is still scope for error and the calculation will not be very exact. However, if you take readings regularly over a long period you will have an idea of seasonal flows and can estimate how much reserve capacity the pond will need.

The *jug method* for measuring the output of a spring is even less accurate but a useful estimate. Put a measuring jug or bucket to catch all the spring water (difficult in a stream but possible with a little ingenuity) and see how long it takes to fill it to a specific level — without letting half the water splash straight out! This can be used for a crude estimation but only up to perhaps 5 or 10 litres (1–2 gallons) per second at the most.

The *thin-plate weir method* makes use of a sheet of wood or metal with a notch cut out of its upper edge, either a V-notch for flows up to about 100 litres (22 gallons) per second or, for faster rates, a rectangular one. The principle is that the channelled stream water accumulates against the weir's upstream surface and the rate of flow of the water over the weir depends on the height of the accumulated water. The height of the weir must ensure an adequate drop between water levels above and below the plate and should be at least 75cm (30in) above the downstream level, with at least 60cm (24in) depth of clear water against the upstream face (less for a V-notch plate). The depth of water passing through the notch is governed by the volume of the water.

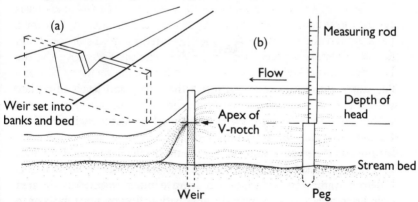

Figure 3.3 **V-Notch weir to gauge flow**
(a) *Place the weir across the stream, set into banks and bed to intercept all the stream-flow. Drive a peg into the bed about a metre (3ft) upstream, with its top at exactly the same level as the angle of the V-notch.*
(b) *Set the zero of the measuring rod at the top of the peg and measure the depth of head of water above the peg*

The V-notch must be cut to exactly 90 degrees, with its sides set at 45 degrees from the horizontal, and a rectangular notch must have its base horizontal. In both cases, the edges of the notch must be straight, clean and sharp for accuracy, so a thin sheet is appropriate, or on thicker sheeting the edges need to be bevelled on the downstream face at an angle of 45 degrees. A rectangular notch is ideally about two-thirds the width of the stream, and a minimum of 30cm (12in) across. V-notches need to be 15cm (6in) deep for a flow of up to 45,000 litres (10,000 gallons) per hour, or 30cm (12in) up to 225,000 litres (50,000 gallons).

Drive a measuring peg into the streambed about a metre (3ft) upstream from the weir (which must completely traverse the stream, either set into the banks or set into a dam so that all the flow has to pass over the weir). The top of the peg should be exactly level with the angle of the V-notch or the crest of the rectangle. Now measure the height of water above the peg: this 'head' represents the depth of water flowing over the weir. Then use the tables in Appendix C to calculate the volume of the flow.

Demands for flow vary, of course. For example, an intensive fish-rearing enterprise might need something like 430,000 litres (94,600 gallons) of pure, fresh water inflow per tonne of fish per day, so access to a clean and reliable supply absolutely dictates the siting of the farm. Watercress beds need perhaps 3–5 million litres per hectare (300,000–500,000 gallons per acre) per day.

Red Tape

The water authorities seek to protect the purity and levels of natural water, and they consider the wider interests of *all* water users, including wildlife. Various Water Acts therefore enable them to control the way in which water is removed (abstracted or impounded) and returned to the natural water system. Full details of the legal situation are given in Rick Brassington's invaluable book, *Finding Water*.

Before water can be abstracted from a main watercourse (which includes rivers and streams), an abstraction licence must be sought from the local water authority (WA). This can take some time and means a lot of questions about the maximum quantities to be abstracted, as it will affect downstream users. A WA impounding licence is required for a pond fed by direct land-drain inflow, or one created by embankments or dams which retain the water above normal ground level. If dams are to be built, it is essential to consult the WA in any event. Dammed reservoirs capable of retaining 25,000 cubic metres (25 million litres or 5.5 million gallons)* above the lowest natural ground level have to be designed and supervised by a qualified, approved, Panel 1 civil engineer, and even in the case of much smaller dams it is wise to bring in a competent engineer in the design stages. It may also be necessary to apply for planning permission from the district council, and such permission certainly has to be sought if a new pond is to be built within 25m (80ft) of a classified road, whatever purpose the pond serves. Large dammed reservoirs have to be registered with the local council and are liable to periodic inspection (at your expense) by county council inspectors.

A simple scrape pond which is allowed to fill with surface run-off or groundwater does not require a WA licence, unless the water is then pumped off for irrigation or other purposes, in which case an abstraction licence is needed.

A new farmland pond could be classified as a change from agricultural use and would in that case need planning permission. However, it remains agricultural if it is used for watering livestock, irrigating crops, reserving water for fire-fighting or is registered with the Ministry of Agriculture as a fishfarm (maximum 2ha/5 acres; minimum 0.8ha/2 acres for trout, or 0.2ha/0.5 acres for coarse fish). An angling enterprise is *not* a fishfarm and to qualify for Ministry registration the fish must be kept live with a view to selling them or transferring them to other waters. If fishing rights are sold, the pond use is classified as recreational, not agricultural. Some fishfarms do require planning permission and they will also attract the WA's attention, because any effluent produced will be classified as *trade* effluent and is subject to WA control. The same applies to any enterprise where there is commercial gain, unless there is no inlet or outlet of water to and from the pond. Angling, however, whether stocked for private use or for others, does not usually come into the 'trade effluent' category, though there is a fine dividing line: could anglers be said to be paying for the *fish* as opposed to the fishing? If the angling waters are let, or perhaps run by a syndicate, planning permission will be needed.

The WA also has a duty to maintain, improve and develop freshwater fisheries in its administrative area and there are necessarily very strict regulations concerning the health of fish. Indiscriminate fish move-

*e.g. a lake 1.25ha × 2m deep, or 3 acres × 6ft deep.

ments can spread fish disease and the authorities have been made responsible for supervising transfers: live fish from any source may not be transferred into a pond, new or old, unless the WA has given permission for the movement. Contact its fisheries officer, who will inspect the transfer. It may also be necessary to have Ministry permission again in the interests of curtailing the spread of fish diseases, and the fisheries officer will advise where this is necessary. The Ministry contact will normally be the divisional surveyor of the Land and Water Service.

Non-toxic spoil from a pond excavation can be used to infill another pond or pit on the same agricultural holding, but planning permission is required if the infill is transferred from a different holding. Planning permission is also required if there is any mineral extraction during the course of pond construction, though in the case of an agricultural fishpond this only applies if the depth of the pond is more than 2.5m (8ft) or its area more than 0.2ha (0.5 acres).

Any works (even planting a tree) carried out within a few metres of what the WA calls a 'main river' requires consultation with the authority, because they must retain unobstructed access to the river for maintenance, etc. In fact, whenever an activity affects any watercourse in any way, it is advisable to have at least an informal discussion with the WA first. Apart from your statutory obligations, you will find that the staff can give plenty of friendly and well-qualified advice on all aspects of water management.

Herbicides are considered a 'controlled form of pollution'. If a pond owner can guarantee that the herbicide sprayed on enclosed water cannot escape to other waters by outlets (constructed or accidental, including seepage) the WA will probably not become involved, but if a watercourse is to be sprayed, or a sprayed pond flows into a watercourse, then advance permission must be obtained from the WA, and all downstream users must be given due warning. There are strict regulations about acceptable herbicides and methods of application and special care needs to be taken if, say, a glasshouse business is abstracting water downstream for irrigation.

Damming or diverting a watercourse for water *power* is an impoundment but, as long as the water is returned to the main stream substantially unchanged in volume and quantity, there is no abstraction.

Abstraction of groundwater needs prior consent from the WA — for example, new wells and boreholes. The authority will require detailed large-scale plans of location, depth, quantities to be abstracted and area of catchment.

Also check your own WA's local bylaws, particularly where a watercourse is involved. 'Watercourse' can include rivers, streams, ditches, drains, dykes, sluices, sewers, and even dry channels which could carry water in certain circumstances.

Waterfowl which are not native species may not be released onto a pond in the wild unless they are flightless — either pinioned or wing-clipped.

Advice and Funds

Before starting any site work, it may well be worth applying to one or two organisations not just for advice but also for possible grants. For example, the Countryside Commission might give a grant for pond restoration if it is of benefit to the general public, which means it must be visible from a public footpath or road. The Ministry of Agriculture might give grants as part of the Farm Improvement Scheme for livestock watering or fire-fighting reservoirs, but not for fishing. The Nature Conservancy Council would only give grants in very special cases — perhaps for the management of ponds particularly rich in wildlife. It is also worth trying major commercial concerns like oil companies, especially if they have perhaps been test-drilling in the area and causing a little consternation among the local people so are in need of some good PR. The EEC might be another source of funds; for example, 1987/8 was European Year of the Environment, and there was scope for project grants for improving conservation ponds. Talk to your local WA about grants, just in case, and to the National Park Authority where relevant. Also try local government: county councils have countryside officers who will always give general advice on conservation ponds and the council does sometimes give grants for interesting projects.

For advice (as opposed to funds) the other initial sources are, where appropriate, FWAG (for farmers, landowners and land managers), the Ministry's ADAS (again, for farmers and small-holders), the local authority's planning department (useful to have informal discussions when the pond or enterprise is no more than a vague dream, because they can advise you about unexpected planning regulations), BTCV (whose interest is conservation, of course, and who might even find you a workforce of volunteers), the district civil engineer or divisional surveyor, the WA's fisheries officer (who can advise on any aspect of water management where fish are concerned), the Game Conservancy or the British Association for Shooting and Conservation (wildfowl), the Royal Society for the Protection of Birds. . . indeed the list is endless! Some addresses are given in Appendix E.

Site Survey

It is *essential* to undertake a proper survey of the topography of the site in order to plan excavations, even for a small-scale pond. To estimate final water levels by eye alone is foolhardy: the contours could be very deceptive and the result will be water where it is not wanted.

A simple amateur tool is the 'dumpy' level, which is a horizontal telescope on a pivot and tripod. It incorporates a spirit level and siting cross-hairs. The telescope needs to be set precisely horizontal, as indicated by the spirit level, and its cross-hairs are used to 'read' a calibrated, extensible staff which is usually rested on top of pegs set at various points on the site. Set up the dumpy in a position from which most of the area can be surveyed without having to re-site the tripod. Drive a series of posts around and across the site and use the dumpy to align a consistent horizontal level which can be marked on each post to indicate the desired water level. Then drop the sights to the levels of the proposed contours within the pond and mark these as well. A dumpy level can be borrowed from a local field centre (if relationships are good) or one can be devised with a cardboard tube, cotton for cross-hairs and a little ingenuity. There are other methods and equipment which are either more or less sophisticated.

To measure slopes very simply, drive a stake in at the highest point and another some way downhill. Tie a string between the stakes. Attach a spirit level to the string, then slide the string up or down each stake until it is shown as horizontal. Measure the height between the ground and the string on each post. The slope is the difference between the two measurements, divided by the distance between the stakes, and if the latter is a hundred measuring units, then the slope is in terms of percentage. This method can be used to stake out the heights of pond embankments.

If the construction includes a dam to impound the watercourse, you must consult the local WA who will advise on levels and possible consequences. Seek professional assistance with dam-building: it is not a task for amateurs, either in concept or execution.

With banked ponds, especially those used for fish, the pond bottom should have something of a gradient so that it can be drained when necessary, and the strongest embankment is at the deepest end. If fish are to be farmed, make sure the pond site is high enough to enable it to be drained regularly: built it up with embankments or dams if necessary.

Draw up a detailed plan of the pond, whatever its ultimate purpose, and design it bearing in mind the points made in the various POND BOXES which suggest some ideals for purposeful ponds. Make the pond as large as possible; the bigger it is, the more stable an environment it will offer to pondlife.

Farm Ponds

For Livestock

Make sure the water source is not polluted (test regularly) and is continuous and clean.

Do not allow cattle actually into the pond as they will soon churn up the edges and bottom. Make a fenced-off drinking bay with concrete or hardcore ramp.

Have a reasonable slope to the access so that natural drainage reduces excessive poaching.

For Fire-fighting

Site within 100–150m (330–490ft) of buildings, and have hard-surfaced roads and bankside standing for access by firetenders.

Minimum capacity 20,000 litres (4,400 gallons) up to 90,000 litres (200,000 gallons) (for example, 10 × 10 × 1m (30 × 30 × 3ft) deep). Have sump or deep water near the access bank.

Except for the hard-standing access areas, the outlines and banking can be irregular and fringed with marginal plants. Keep most of the water surface free of plants: avoid invasive marginals and aquatic floaters — deter them by dropping banks quite sharply beyond the planting shelves.

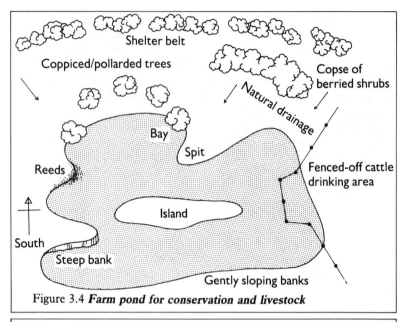

Figure 3.4 **Farm pond for conservation and livestock**

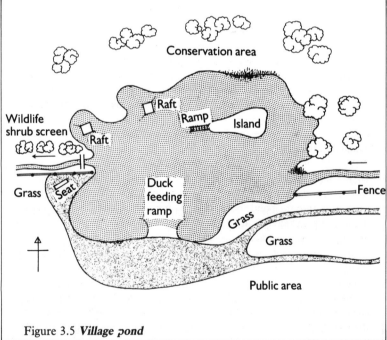

Figure 3.5 **Village pond**

Conservation Ponds

Variety is the key: irregular shapes; a range of depths (deeper for fish refuge, shallows for feeding and wildfowl dabbling); some steep banks and some very gentle; islands, bay and spits; wildfowl nesting rafts.

Full sunlight from the south, with a few carefully controlled trees for shade here and there elsewhere, but allow ample light.

Leave a shelter-belt of trees set well back from the water, beyond reach of their shade and leaf-fall.

Offer a variety of habitats within and beside the pond: aquatic weeds (step or shelve banks for wider range of species), areas of deeper, clear water; marginal plants and reed beds; copse of berry-bearing shrubs.

Buffer pond from agricultural run-offs (fertilisers, sprays) and provide seclusion and peace for wildlife.

Link site to other habitats — provide hedgerows, ditches, highways of rough grass, etc. to link with old meadowland, marsh areas, different ponds, broadleaved woodland, copses.

Village Ponds

Clear southern aspect for maximum sunshine.

Plenty of variety for wildlife, as described for conservation ponds.

Balance between access by people and privacy for wildlife: try and keep part of the pond private for the latter. Islands very useful in this context.

Duck-feeding areas, with paved ramps to avoid ducks trampling heavily used areas into a quagmire.

Projecting piers for favourite places for observing pondlife, to avoid damage to bank.

BEWARE OF SUBURBAN TIDINESS, which is out of keeping with a village pond. Any 'furniture' should be in natural materials, e.g. stumps for seats. Avoid asphalt (could have problems with run-off of toxic substances into the pond) but need at least a few stepping stones on grass to protect from heavy wear and tear. Keep motor vehicles well away — to avoid pollution (oil, etc.) as well as for safety and peace.

Plant with native species for preference, including marginals and shrubs — there is plenty of choice for attractive flowering plants or interesting foliage.

Provide discreet litterbins (making sure they are emptied regularly) and safety equipment (ladder, rope with handgrip, long pole, firebrooms).

Fence off dangerous areas, e.g. near weirs.

Duckponds

Wild Ducks

Wildfowl need access to larger expanses of open water for safe landing and roosting, gentle weather-protected loafing banks for preening and resting, lack of disturbance, clear flight lines unobstructed by trees but ample cover for nesting and shelter, and a good variety of aquatic plants. Not too close to tall hedges or trees. Ducks will also be shy of high banks because they cannot see approaching danger and because they need shallow slopes for easy access to loafing areas.

Give very varied shoreline with plenty of secluded bays (or cut some inlets in reed beds). Drakes are territorial in the mating season and need own area out of sight of neighbours.

Reasonable size of pond is minimum 20m (65ft) across, more than 1m (3ft) deep in centre, 0.3m (1ft) in shallows at least a metre (3ft) wide for dibbling. Deeper pond needed if source is surface run-off rather than running water: say 2m (6–7ft) at centre for an area of clear water, with rest of pond 0.5–1.2m (2–4ft) and shallower margins.

Give shelter from high winds with windbreak set well back and plenty of fringe cover and weedy shallows. Also ample cover near loafing banks: ducks will not fly for a few weeks after moult and feel very vulnerable then. Banks with bramble patches for food and shelter.

Good aquatic plant population will give food and also harbour small creatures for ducks (snails, crustaceans, etc.). Note that ducks and fish are compatible competitors for food! Ducks will appreciate barley grains scattered on shallows (away from banks to deter rodents) or unthreshed sheaves suspended from stakes with heads just below water.

Ornamental Waterfowl

Minimum 4m (13ft) diameter. Maximum depth 0.75m (30in) at centre, shallows 0.3m (1ft). Good supply of aquatic plants, weed, freshwater snails, crustaceans. Wings of non-native species must be clipped or pinioned. No housing necessary but shelter will be appreciated.

Swan Pond

Need ample space — plenty of open water and no runway obstructions on the pond or its surroundings. Ample aquatic plants, especially submerged species. Can be aggressive when nesting — anglers beware!

Domestic Ducks

Only need enough water for paddling, washing, preening. Pond could be 2m (6ft) in diameter and shallow, but would need more than one pond, use in rotation to avoid build-up of disease. Accustomed to having access to dry housing, especially in cold weather. Interbreed readily with wild mallard.

Trout Ponds

Clean gravel bed with fair flow of water for spawning, average 0.15–0.6m (6in–2ft) deep.
Oxygen vital — running water, cool temperatures, submerged aquatics for oxygen and cover.
Higher pH values for better growth (minimum 6.5, up to 8.4): add lime if necessary (ground limestone on pond bottom could last four years).

Trout Lake: average 1.5–2m (5–6ft) deep over at least one-third of area with pockets 2–2.5m (6–8ft), rest shelving to 0.6m (2ft) at banks. Continuous waterflow (350,000–550,000 litres or 80,000–120,000 gallons per day). High calcium input. Leave new lake at least one season after filling and settling before introducing fish. Screened inlets/outlets. Natural vegetation from feeder stream. High alkalinity gives continuous algae and duckweed.

Stocking: Up to 50 large fish (0.5–1kg, 1–2lb) in 0.2ha (0.5 acre), or 100 at 15–30cm (6–12in), or up to 250 yearlings. If mixed, maximum 150 fish unless fed artificially.

Trout Stream: Deep lying-up pools dug out up to 1.2m (4ft) deep. Rainbows will not spawn; brownies might in cold shallow water on gravel or chalk bed near source of spring. Encourage watercress, etc.

Mixed Stocking: 1.85 sq m (24 sq in) of surface area of water per fish 2.5cm (1in). Could also include carp, tench, rudd, gudgeon and a few roach but ensure that trout find enough food. Trout are very territorial between themselves.

Earthponds for Fishfarms

Bottom: Smooth and compact, then dig drainage ditches — principal ditch along full length of pond to outlet with its base 50cm (20in) wide, sides sloping 1:1.5, minimum fall 1 per thousand (2 or 3 per thousand gives better drainage); herringbone ditches from sides to central ditch, falls 5 per thousand; collection basin in front of outfall sunk below level of pond bottom and lined with concrete for firm base during fish collection.

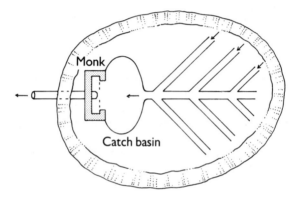

Figure 3.6 *Earthpond drainage layout*

Inlets: Controlled with on/off supply; filter traps to keep out unwanted wild eggs, etc.; mesh to keep fish in. Trout cultivation 5 litres/second/hectare (intensive 100 litres) (1 gallon/second/2.5 acres (intensive 22 gallons)). Control inflow to adjust temperature and oxygen (mechanical aeration ideal): salmon need 9mg oxygen per litre, carp 6mg.

Outlets: Monk with fish screen, to control water level and enable draining off of bottom water when it is necessary to remove poor-quality (deoxygenated) layer.

Management: Drain after each harvest, allow bottom to dry, remove excess mud and debris, plough the soil, apply lime to counteract acidity, disease and parasites, manure to maintain fertile layer of soil. Could use direct-manure system: 250 ducks per hectare (2.5 acres) of pond, or 200 pigs per hectare (2.5 acres) of pond beside the pond (manure washed in). Need balance between fertility and fouling.

Angling Ponds

Fish need well oxygenated water with ample pools of deep water (minimum 2m (6ft)) and shady areas for refuge.

Trout need running water and gravel-bedded spawning shallows.

Stability of temperature is important; therefore the bigger the pond the better.

Vegetation important for shelter, for underwater oxygen and as habitat for invertebrates on which fish feed. As well as aquatics have a few overhanging bankside branches.

Do not stock with fish until a new pond has settled and established its plantlife and natural food (at least a year). Aim for plenty of variety of plants.

Achieve maximum shoreline by creating inlets, promontories and islands. Anglers need some steep banks or piers for deep-water access and also need privacy provided by well-screened bays. Keep angling stations clear of branches which might entangle cast lines. Provide access for anglers but at the same time keep the pond and fish secure from poachers!

Discuss stocking and management with WA fisheries officer.

Consider making a complex of ponds for different angling requirements and to include conservation.

Watercress Beds

Very clean water (e.g. chalk spring) at constant temperature, slow and steady current. Best source is stream over chalk, limestone or possibly greensand or light red sandstone. Avoid peaty acid soils.

Shallow water, up to 0.3m (1ft). Establish plants in shallows at sides to begin with.

Intensive culture in purpose-built concrete-sided terraces, each bed with separate inflow control direct from feeder stream. Must be able to drain beds completely and regularly. Not for amateurs: requires constant monitoring, labour, high capital investment.

Amphibian Ponds

Minimum 4 × 2m (12 × 6ft) — larger for great crested newt or for toads. At least part of pond down to about 0.5m (say 18–24in) to prevent complete freezing in winter.
Variety of shape and depth.

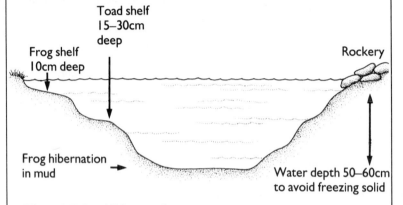

Figure 3.7 *Amphibian pond*

Frogs: broad shallows about 10cm (4in) deep for breeding.

Toads: shallows 15–30cm (6–12in).

Easy slope on grassy bank for exit of emerging young, with marginals for cover.
Area of open water plus ample native plant species, especially submerged and emergent aquatics. Broadleaved species like starwort and water soldier for newt egg-laying. Toads twine spawn strings around submerged parts of plants.
Rockery at pondside useful for hibernation, sunbathing, shady retreat and protection.

NO FISH: they will devour all the tadpoles.

Chapter 4

CONSTRUCTION

My skill goes beyond the depths of a pond. . .
Martin Parker: **Upon Defacing of Whitehall**

Dams

The design and construction of a dam is best left to an expert engineer, but here are some basic principles.

Dams are *not* designed to have water flung over them; due allowance must be made for overflow systems to divert water away from the dam long before the water level rises over its crest and weakens its structure. It is important to have an adequate stormwater spillway to divert flood waters away from the dam and remove stress at times of spate. (See Outlets). This is, of course, in addition to the main spillway which takes the daily flow.

Dams are under constant pressure from the weight of retained water (much more than from the current) and they need to be very strong and completely watertight, especially around the edges where leakage is most likely and which will, in due course, weaken and undermine the entire structure. The weakest point is where the weight of accumulated water is greatest — at the base of the dam. Make sure there is not an unsuspected pervious layer under the dam site: check it with a test pit during the planning stages.

The dimensions and shape of a dam should be checked with the local WA in the planning stages. Their criteria will probably state a freeboard of perhaps a metre (3ft) above the highest anticipated water level, with an emergency spillway level at 30cm (1ft) above normal. The minimum width of the top of the dam (its crest) should be 2.5m (8ft) and the slope of the upstream face should be no steeper than 1 unit vertical to 2.5 units horizontal, preferably 1:3 or 1:4. The slope of the downstream face relates to the height of the dam: for example if the height is 4m (12ft) the slope should be 1:2, or 1:2.5 for a dam 5.5m (18ft) high, or 1:3 for an 8m (25ft) dam.

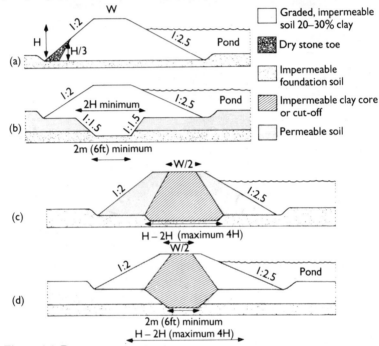

Figure 4.1 **Dams**
(a) Homogeneous dam on impermeable foundation soil.
(b) Homogeneous dam on permeable soil, with cut-off wall.
(c) Clay-cored dam on impermeable soil with keyed foundation.
(d) Clay-cored dam on permeable soil with keyed cut-off [Based on the MAFF booklet Water for Irrigation]

The width of the dam is proportional to its height, and the height is related to capacity. In general, earth dams are built to about 1.5m (5ft) higher than maximum water level to be on the safe side. A minimum width at the base should be four times the height of the dam, and the crest width should be at least half the height measurement. If soil quality is doubtful, increase the base width to six times the height, and the same applies if the dam is more than 6m (20ft) high (at which point it is way beyond the capabilities of the amateur). The width of the crest should never be less than a metre (3ft), however low the dam or embankment.

The core of the dam must be impermeable and the ideal material is thoroughly sticky clay. However, pure clay shrinks and cracks if it is allowed to dry out at any stage, and it must be mixed with sand, gravel or loam (but avoid those containing humus, which rots and creates structural weaknesses). For a clay core the clay content of the mixture should be at least 30 per cent, and for a homogeneous dam (i.e. of impermeable soil throughout) the clay content should be less than 30 per cent but more than 20 per cent.

Clearly it makes sense to use the site subsoil for the dam if it is at all suitable: it will save considerably on costs and the excavation of the dam material creates a 'borrow pit' which forms part of the actual pond.

To get an idea of the clay content:

(a) Roll a small sample of subsoil between your hands. If it falls to pieces, there is not enough clay. The ideal mixture becomes a little tacky when it is moist and the clay content stains your palms. For a clay core the sample should feel really quite sticky. Roll the sample into a ball and let it dry naturally to see whether it has a tendency to crack.

(b) Alternatively, sieve the sample into a glass container, add water and a spoonful of kitchen salt, and mix by shaking. Let it stand for a day or so until it settles into bands of sand, silt and clay, in that order from the bottom up. For a clay core at least a third of the settlement should be the clay layer at the top. For a homogeneous dam, with a 20–30 per cent clay content, the mixture should feel a little gritty.

Remove any stones from the clay mixture before it is used, and keep it moist at all stages of construction to avoid cracking (cover with topsoil as soon as the claywork is complete).

Prepare the site by clearing vegetation and roots, setting the topsoil aside for topping off the dam and embankments. Divert the stream, drain the site and peg out the dam outline. Dig out foundation trenches down to something solid, extending them sideways to key into the banks on either side of the dam.

The point of a clay core is not just to prevent water seeping through the dam but also to deter it from finding its way underneath, which it will certainly try to do. For that reason the core needs to be sunk at least a metre (3ft) deep into the pondbed subsoil, even if that soil has a high clay content, and if the subsoil has less than about 30 per cent clay content the foundation needs to be extended further down to be keyed in place with clay trenches another metre (3ft) deep. In general, the foundations should go down into the bed to a depth equivalent to that of the depth of water to be impounded. The foundation ditches (sometimes called the 'cut-off wall') should be at least 1.2m (4ft) wide at their base and their sides should be well sloped.

Build up the clay core gradually in layers no more than 15–20cm thick (say 6in at a time) and compact each layer thoroughly before introducing more clay. Compaction of bigger dams is best done mechanically with rollers, tamps or tractor wheels where space permits, but with many ponds, where the scale is more human, the compaction is best carried out by trampling feet (human, bovine, ovine, or asinine), in which case the layers should be thinner — perhaps 10–20cm (4–8in). Thorough consolidation is essential at each stage and for best results it is necessary to moisten the clay, but only just enough for it to form into a ball when rolled in the hand.

Once the foundations have been laid, the main body of the core is built up in the same careful way, layer by layer across the site, and gradually bonded into the adjacent banks. Pay special attention to the bonding at pond-bottom level.

The final layers and shaping can be of the local subsoil, even if it has less than 30 per cent clay (as long as the core is impermeable), and this, too, should be carefully compacted. The top of the clay core, of course, should be above the highest anticipated water level and allowances should be made for settlement during or after construction — say 10–20 per cent after compaction.

The crest should be slightly sloped or cambered for rainwater run-off. The pond face can be lined with stone to protect it from wave erosion if the pond is of any size. Some people plant reeds for the same reason, but they will act as a silt trap and would be more useful elsewhere.

Spread topsoil over the crest and on the downstream face, then sow it with grass seed or turf it. The grass helps to consolidate the dam and is the only suitable cladding; it can be kept mown or grazed by light-footed sheep or by geese. On no account allow trees or shrubs to grow on any part of a dam. Even smaller plants should be discouraged: it is really only grasses which will create the ideal mat, holding the top 2.5cm (1in) of soil firmly in a web of fine roots. Roots which are any more substantial will create seepage channels.

Apart from controlling the grass, future maintenance of a new dam is largely concerned with careful monitoring for seepage. Clay tends to 'sweat' during the first year anyway, so there may be no need to panic, but if there are damp patches they will probably increase. If the dam was built by contractors, they should be required to check the dam and make good where necessary. Minor leaks in small dams can be sealed with a mixture of ground-up ashes and aluminium silicate, which swells on contact with water: empty a bagful or two in the general area of the leak. Always act sooner rather than later if a leak is suspected, because once water has found an escape route it will work on it and one day the dam will burst.

On a rock bottom which is truly impervious and solid, the dam could be made of rock with a core of clay or concrete, or could be rockfill with the pond-side slope faced with impervious clay-based soil. Make a retaining wall or revetment of stone, brick or concrete (wood and steel pilings may not be strong enough) and reinforce the pond site of the dam with a spoil bank sloping from a wide base.

Long, low dams are cheaper and safer to build than high ones. Small dams no more than knee or calf height could be made of assorted rubbish (turves, old doors, pieces of sheet metal or compacted soil) set into a trench and protected with a seal of puddled clay. Even a barrier of sandbags, supported by stakes and clad with corrugated iron, could be used to build up a head of a metre or two (3–6ft) in a knee-high stream — temporarily. Old railway sleepers can be used to dam small streams. But however small the stream, and however small the dam, your impoundment will have effects downstream and you must first of all consult the WA. In theory, even a line of stepping stones interferes with the watercourse and should not be placed without permission.

If the enterprise includes a trout stream, do not be tempted to dam it, which will only cause silting. It is much better to dig scrapes into the stream bed, creating deeper pools in the gravel which will be much appreciated by the trout.

Embankments

Embankments, or dykes, are in effect continuous dams encircling an off-stream pond (a dam impounds water on-stream) to retain water above the natural ground level. Building techniques are similar to those for dams except that in most cases the water pressure will not be as severe. As with dams, there should be ample freeboard between high-water level and the crest of the banks, and a depth of 2m (6ft) of water needs a freeboard of at least 30cm (1ft), increasing to 50cm (20in) for deeper ponds. Allow for settlement during and after construction.

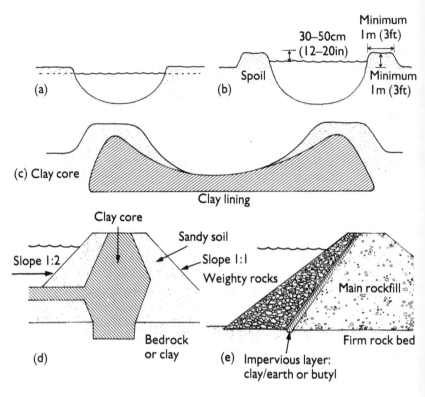

Figure 4.2 **Embanked ponds**

(a) *Simple scrape pond with no embankments, dug down to below the level of the watertable.*

(b) *Embankments increase the pond's holding capacity by allowing the water to rise above the surrounding ground level. There should be a minimum of 30cm (12in) (50cm (20in) for a larger pond) between the highest expected water level in the pond and the top of its embankment, and the construction should be at least 1m (3ft) wide at the top and at least 1m (3ft) higher than the surrounding ground level.*

(c) *Where the watertable is low, the clay bottom lining should be extended to form a continuous impermeable core in the embankments.*

(d) *If sandy soil is used for an embankment, it needs a clay core keyed well down into the underlying bedrock or clay.*

(e) *On a firm rock bed, embankments can be of rockfill with an impervious internal layer of clayey soil (or butyl if the banks are small) faced with heavy rocks on the pond side. A concrete weir or a rock dam with a concrete core might be more suitable*

Embankments are most economically made in 'cut-and-fill' situations, where soil excavated from the pond is used to form the barriers.

Leave a shelf about 2m (6ft) wide between the pond excavation and the banks to make future maintenance work more accessible. The same criteria for consolidation of each layer apply and it is essential that the embankments are watertight and stable under all conditions. Slopes and proportions are as for dams.

Embanked ponds are economical to build in that the volume of the water (if that has a value) relates favourably to the volume of soil excavated: there is a greater depth of water per surface area of excavation than there would be for a pond with its water level near natural ground level. However, unless the embankments are well built and maintained, there are more likely to be problems with water loss through seepage.

On permeable soils the embankments need a central impermeable core 50cm (20in) thick, keyed down to well below the level of the pond bottom and rising above maximum water levels. Lined ponds can benefit if the embankment cores link in tightly with the clay lining.

Wave erosion, especially in the case of new embankments, can be a serious problem. Try protecting the bank with a floating log-boom (6m (20ft) long, 25–30cm (10–12in) in diameter, held in place with stakes but able to rise and fall with the water level). In a small pond, use heavy grass sods as facing on dams and banks. Or pile up some fibre sacks filled with soil and backfill the wall with rubble. The fibre will eventually rot away, by which time the soil will have been consolidated by plant growth. Or weave wattle fencing as protection: if it is made of willow it will probably sprout and give additional shelter for wildlife.

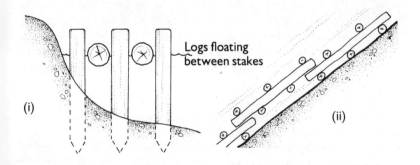

Figure 4.3 **Bank booms**
Logs floating between stakes rise and fall with the waterlevel and protect banks from wave erosion.
(i) Side view; (ii) top view

Outlets

All ponds need an outlet of some kind, whether it is in regular use to allow water levels to be controlled, or for occasional emergency use in times of flood, or for management purposes if the pond needs to be drained from time to time. In many off-stream ponds the daily control of water levels is exercised at the inlet, and it is only necessary to have an emergency overflow system and a plugged bottom outlet for draining during maintenance.

The basic types of outlet are weirs and sluices (also used to control inlets), funnels and plugs (for bottom draining), and spillways.

Spillways

Spillways are essential where a stream's flow is interrupted by a dam. The natural flow of the water must be able to continue without damaging the barrier and certainly without overtopping it. Spillways allow an outflow over their tops instead, and are generally set to one side of the dam. There should be a primary spillway or spillways for the usual daily outflow, and at least one stormwater spillway to act like a bath overflow in times of flood.

The combined spillways must have the capacity to deal with extreme situations and ensure that at all times the pond level remains well below the crest of the dam or embankments. There should be at least 1.5m (5ft) freeboard between the level of the spillway floor and that of the crests, or the water level in the spillway should be at least 1m (3ft) below the crest. The spillway should then fall away with a slope of, say, 1:75 and should discharge the flow at an outfall well below the embankment. Its sides and floor will need protection from erosion by the constant friction of running water. Some spillways are built with a concrete sill across them in line with the adjoining embankments to prevent the outflow water backing up into the pond.

The wider the spillway, the better it will cope with high flows, and the bigger the catchment area, the bigger the spillway needs to be. Always build it much wider than seems necessary, and built it solidly. Take into account that the rate of surface water run-off from the catchment area will be affected by topography, plant cover and soil porosity. Rain from a torrential storm will be far more dangerous in a hilly catchment area with poor vegetation cover than in a gently sloped area of the same acreage but well planted.

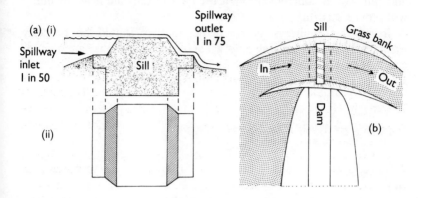

Figure 4.4 **Spillways**
(*a*) *Spillway sill to prevent backing up of water. Note gradients of spillway floor on each side of sill.*
    (*i*) *Side view; (ii) top view*
(*b*) *The sill should be aligned with the crest of the dam. It can be made of concrete, brick or rock*

To calculate the approximate size of the spillway, find the level of highest flood on the stream's banks (look for deposits of debris), measure the width from bank to bank at that level and calculate the average depth that produced the floodline. These figures represent a cross-section of the stream in flood. For a small stream, divide the flood depth by the desired depth of flow, then multiply by the highwater width. The result is the necessary width of the spillway.

For a catchment area of more than 20ha (49 acres), simply divide the area by 2 to give an arbitrary spillway width in metres (yards), then add a 3m (1yd) safety margin. If the area is more than 50ha (125 acres), discuss the matter with the local WA fisheries officer who will be able to find information on the amount of run-off from the drainage area.

Where the main course of the stream naturally bypasses the pond, with a controlled inlet to it, such large spillways are unnecessary: the capacity need only exceed the capacity of the inflow, and the main flood control is by means of closing off the inlet.

Spillways are usually built at either or both ends of a dam, preferably in such a way that the impounded water is funnelled towards them along a curved approach, and they must be paved far enough along to carry the water well away from the dam. Concrete spillway channels can be built on concrete supports for greater stability, with the first support extended for extra strength at the potential weak-point between dam and spillway. A concrete apron at the back of the dam near the outlet will reduce scouring.

Stormwater spillways as emergency outlets should be properly maintained. It is all too easy to forget about them until they are needed, and they must be kept clear in case of flooding. They must be dug into solid ground and can be grassed as they are rarely in use.

Funnels or Weir Chambers

A funnel outflow system is often used for larger bodies of water with wide catchment areas or in areas of high annual rainfall. It is basically a vertical pipe or shaft in the pond near the dam, with its top at the normal water level. Overflow water travels down the pipe to a chamber (which also contains a plug to enable the pond to be drained for maintenance) and thence sideways along a sloping pipe built under the dam just beneath the pond bottom. The outflow falls onto a sloped slab and thence to the main stream. Emergency overflow pipes or spillways must also be installed for stormwater.

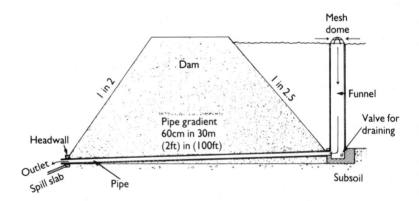

Figure 4.5 **Funnel**
Plan showing siting of pond's overflow funnel and outlet pipe

The funnel needs to be capped with a mesh cage to filter out debris and to keep fish from escaping, and the filter needs regular clearing. Make sure the cage is a dome, not a cup which could trap ducklings and would anyway soon clog with debris. The shaft can be made of brick, or of concrete pipe rings perhaps 90cm (3ft) in diameter for a pond of 1ha (2.5 acres). All outflow pipes, whether funnels or emergency overflows, should always be of much larger diameter than might be thought necessary so that, like spillways, they can cope with extreme situations, and this applies whether the pond is a large estate lake or a small garden pool.

There are obvious weak points in the system, particularly where pipes run under the dam or meet the pond bottom, and due precautions should be taken to monitor for leaks.

Drains

It is important to be able to drain *any* pond completely from time to time, especially if it is stocked with fish or if there should ever be a need to carry out maintenance work (there will be!). If the deepest point is lower than any normal outflow system, or if a plugged drain does not form part of a funnel outflow, then a separate pond-bottom outlet can be installed. It could be a pipe through the base of the dam (not less than 15cm (6in) in diameter) leading from a concrete or brick chamber in the pond protected from debris, algae and fish by a screen and set above the level of the bed to keep clear of silt. The outlet has a plug, either in the pond (in which case a long iron key will be needed to locate and turn the valve) or on the downstream side of the dam where the pipe issues. Because memory can blur, it is important to have some means of marking the site of the plug chamber: it soon becomes invisible under silt and vegetation. The easiest is to mark the dam wall (along the lines of a treasure hunt: go out x paces and down y paces) unless a permanent rod and wheel are installed.

Marilyn Chakroff, of the Peace Corps, has some pleasingly simple and ingenious ideas for plugged drains in her book *Freshwater Fish Pond Culture and Management*, written for small enterprises in developing countries and full of very useful, sensible advice. Some of them are illustrated here. Note that in fishponds it is important to be able to remove the *lower* water layers, which are more likely to become deoxygenated.

Naturally the outfall from any outlet pipe must be lower than the deepest level of the pond. Beware of seepage between the outside of the pipe and the surrounding soil: lay it in a special trench and fit it with anti-seepage collars, and check regularly for leaks.

The 'plug' system is rather basic, and a more efficient method was invented several centuries ago by those great fishfarmers, the monasteries. It is, of course, called the *monk* and it is described under Weirs and Sluices.

It may prove easier to pump out a pond than to pull the plug, or perhaps siphon out the water (see Pumps, pp. 85–7).

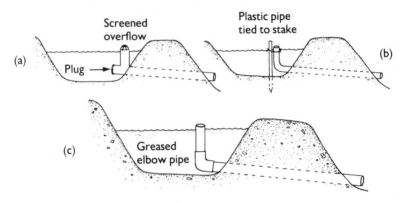

Figure 4.6 **Pond drainage**
(a) *Simple T-pipe with screened overflow at normal top waterlevel and plug for removal when pond needs complete draining.*
(b) *Flexible pipe tied to stake so that its mouth is well above normal waterlevel. To drain the pond, untie the pipe and let it lie on the bottom of the pond. The system can be used as an overflow if the pipe is tied at an appropriate level.*
(c) *A slightly more sophisticated version of (b): the flexible pipe is replaced by a greased elbow pipe. The pond is drained by rotating the elbow joint so that the overflow inlet pipe lies horizontally on the pond bottom [Based on Marilyn Chakroff's* Freshwater Fish Pond Culture and Management*]*

Weirs and Sluices

A weir is defined in dictionaries as a dam in that it obstructs a watercourse. It differs radically from a dam because it is specifically designed for water to overflow it, which would be disastrous for a true earth dam.

Weirs can be simple 'spillboards' slotted across a watercourse — wooden sleepers, perhaps, or thick sheet metal or concrete. Embedded in earth banks, and set lower than the banks of the watercourse, they form simple barriers to build up a backwater. Alternatively, the stream flow can be slowed rather than impounded, by using less obstructive weirs through which some water can still flow. Typical obstructions are stones and logs, or perhaps 'gabions' which are containers (originally basketwork) filled with earth or stones. Concrete paving slabs propped up across the flow at an angle of 45 degrees from the stream bed make good temporary weirs to give quiet, scattered pools for trout. More permanent weirs can be made with tree trunks interpacked with stones and clay and faced with boards, or can be built in concrete or stone.

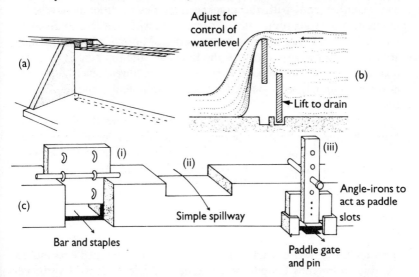

Figure 4.7 *Sluices*
(a) *Typical basic sluice gates: two boards in grooves cut into a wall which is keyed into the banks. The grooves continue in the concrete spill apron to prevent water escaping under the boards. The gates can be interpacked with clay until the pond needs to be drained.*
(b) *One board is raised to determine the waterlevel retained by the sluice. To drain the pond, lift the second board as well so that all the water flows freely under the gates.*
(c) *Some basic sluice designs:*
(i) *Board with staples or brackets for a rod which rests on side walls. The board is set in vertical grooves, with a third groove in the concrete apron, and water normally flows under the raised board.*
(ii) *Simple spillway: lowered section of wall for overflow of water.*
(iii) *Paddle gate fixed to vertical bar with series of holes at different heights for supporting pin. Angle-irons on the downstream face of the wall act as paddle slots, and there is also a shaped groove to accept the base of the paddle and its bar*

A sluice is a system of gates which regulates the flow of water and is in essence easy to manipulate. Sluices can be used to control the inflow to a pond from a stream, or can adjust the water level by regulating the rate at which water leaves the pond at an outflow. Generally, a plate of some kind is raised or lowered in a gap in a wall (perhaps a weir or dam) so that the rate of flow of water escaping under the gate can be increased or decreased. The gate may slide in recesses in the wall, or perhaps in vertical angle-irons set against the outside of the wall. It might have brackets through which a horizontal bar runs and rests on the top of the wall at either end, with the brackets in series at different levels, or might have a strong protruding vertical beam with holes through which the bar passes. Other systems have a series of horizontal boards which are removed or inserted to vary the flow over the top board; if all the boards are removed the pond is reduced to its lowest level. There are also various ratcheted devices for easier manipulation and security of the gates.

A sluice is often under considerable pressure from the force of the water and needs to be strongly built. There is also a lot of force on the far side of a sluice or weir where the water cascades, and the floor here on the 'toe' needs reinforcing with heavy stones to break the fall.

The Monk

This monastic device, said to have been designed by medieval monks for their numerous fishponds, combines waterlevel control with fish management. It is still widely used all over the world for ponds up to 1 or 2m (3–6ft) deep and anything up to 3 or 4ha (7–10 acres) in area. Very often a fishfarm has a series of small ponds, each with its own monk. Historically, fishponds were built in several sizes to accommodate different stages and species of fish, and a collection of such assorted ponds on one site is a clue to their possible origins.

The monk is a cheap, efficient hatch which enables fishfarmers to regulate pond levels without letting fish escape in the outflow. The pond can be drained completely, which is an essential routine in fishfarming where regular cleaning out is important, and where harvesting is often by means of draining.

During initial excavations the pond is given a central drainage channel sloping towards the monk at the pond's deepest point. The monk usually incorporates a catchment chamber for trapping and sorting fish when the pond is drained, or for retaining them while some maintenance operation is carried out. Fishscreens are included in the monk to deter escapers (especially rainbow trout, who are experts) but screens do tend to collect weed and debris and need to be cleared regularly.

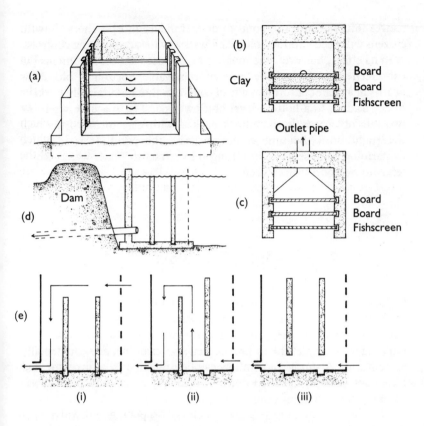

Figure 4.8 **The monk sluice**
(a) *The monk is an enclosed sluice, built of concrete or brick (originally wood). It has three sets of grooves: two for sets of removable boards or blocks with handles and one for an optional fishscreen to prevent the escape of fish from the pond.*
(b) *Top view of the monk. Clay is often packed between the two sets of boards to ensure watertightness.*
(c) *Looking down on the base of the monk: two sets of boards and fishscreen in place, with outlet pipe at the back of the monk at floor level.*
(d) *Siting the monk next to embankment with outlet pipe running through the bank.*
(e) *Choice of board settings:*
(i) *Normal day-to-day position, with water overflowing the tops of both sets.*
(ii) *Position for removing deoxygenated lower layers of water in fishpond.*
(iii) *Position for complete drainage of pond — both sets raised*

A monk is not built into a pond wall like a sluice, though it often touches the wall at the back. Nor is it used as an inlet like a sluice. It can be built beside the dam or embankment, or perhaps a couple of metres (6ft) out into the pond to deter vandals and poachers. Lawful citizens will need a retractable plank walkway for access to the controls.

In its effect, however, the monk is an enclosed sluice. It contains the fishscreen and two parallel sets of removable horizontal blocks or boards in a chamber (often made of concrete, though originally wood or brick) leading to a bottom-level outflow pipe. The space between the two sets of boards is sometimes packed with clay to make the wall watertight until the time comes to drain the pond. It is ideal for fishfarming because it can be arranged to extract the *lower* water, which tends to be poorest in oxygen.

Inlet Controls

Sluices can be used to control a pond's inflow. One of the main problems with stream-fed ponds is invasion by silt, reducing the water depth and blocking the outlets, and this nuisance can be reduced by building a silt trap at the inflow. A weir, for example, acts incidentally as a silt trap. A simple pipe inflow can be protected by a finely meshed filter but it will need constant checking for accumulated debris blocking the incoming water.

The simplest silt trap is a pool made by deepening and widening a section in the stream some way across the inlet so that the rate of flow is reduced and the silt can settle there, or in a walled-off pool at the inlet. It needs to be dug out periodically. It helps if the remaining stretch of stream bed from the trap to the inlet is scraped down to solid matter, dug out to perhaps 20cm (8in) and filled again with gravel, or if a concrete apron is extended from the weir.

There are even simpler methods for small flows. Place a section of large-diameter sewer pipe vertically to intercept the inflow pipe with an outlet on the pond side at the same level, and remember to empty it at intervals. Or put a mesh bag over the mouth of the pipe (but it will soon be blocked with debris). Or place a horizontal screen under the inflow, which should be designed to cascade into the pond through the screen (achieving aeration as well as keeping out the silt). Or take a little more trouble and built a sand-and-gravel filter pit, or make a square concrete box and put it just under the water where a more gentle inflow trickles into the pond. Shovel out the silt now and then.

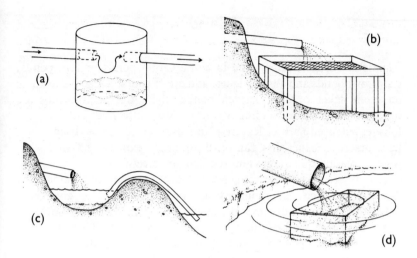

Figure 4.9 **Silt traps**
(a) *Simple system with piece of large-diameter pipe intercepting inlet pipe.*
(b) *Silt screen under inlet pipe.*
(c) *Pool for settling silt. The inflow water then continues to the pond by means of a siphon.*
(d) *Concrete silt tray sunk into pond beneath gentle flow from inlet pipe*

Mechanical Excavations

Peg out the site and give an indication where edges are to have gentle slopes (most excavators leave steep banks). Make quite sure the operator understands clearly what is envisaged. Aim to have the site scraped out rather than dug, and a bulldozer with a curved blade can do the job, working on caterpillar treads if the bottom is firm enough. For larger excavations use a dragline and bucket machine (a scoop and crane with a 15m (50ft) reach, operating from the bank).

Plan the disposal of the excavated material in advance. If the water level is to be contained below the natural ground level, there will be a large amount of spoil to be deposited elsewhere. It may also be necessary to pump the site as the work progresses and the watertable is reached, and you need to have somewhere to dispose of the water as well as the spoil. Many off-stream ponds, however, are designed for a water level above that of the surroundings, confined by embankments created largely from the excavated spoil.

Islands

Islands are of particular value in conservation ponds, village ponds, waterfowl ponds and angling lakes, and are of course attractive features on estate lakes or large garden ponds where trees and shrubs are reflected in the surrounding water — very effective with autumn foliage, coloured bark or feature plants like pampas grass. Islands also have practical value: they can break up larger expanses of windswept water, creating quiet areas and reducing erosive waves.

Islands are essentially refuges and in new ponds they should be designed as such, far enough from the main shores to give birds protection from landbound predators and with adequate cover to camouflage nests and resting places from aerial attack. They also need gentle access slopes here and there, especially for ducklings and amphibians.

There are endless ways of creating islands, which need not even be built from the pond bed upwards. For ducks, for example, floating platforms are in effect islands (see Rafts).

Several predators are quite capable of swimming but will probably decide it is not worth the effort if an island is at least 30m (100ft) from the shore, unless a build-up of silt in the leeside backwaters of the island forms a bridge or sunken causeway, or of course the pond ices over.

Excavated Islands

During pond excavations, leave one or two areas intact to form islands but assess the eventual water level and build them up if necessary while there is still easy access. This can be one answer to trees on the site: rather than the labour of uprooting them, one or two can be left *in situ* surrounded by their own island. Bear in mind, however, that too big a tree may cast too much shade and shed a lot of leaves into the water, or may look out of proportion in the finished scheme, or may die if the water level is too high or too low or if the island's soil is built up well above the original ground level. The banks of a tree island will need reinforcing.

In general, it is better to restrict island vegetation to low shrubs, with an area of shallow grassy banks for access and some marginal plants (shelve the banks if necessary) to give shelter to birds at the water's edge. Fish appreciate some overhanging shrubs for shade and for the insect life that might fall from them. Trout get most of their food in shallows close to the shores, including island shores, so the inclusion of islands in the scheme serves to increase trout feeding areas considerably.

New Islands

All sorts of materials can contribute to the building of a new island — pond-bottom ooze and silt, soil, tree stumps and logs, rubble and other spoil for example. They will not, however, be as stable as excavated ones.

Logs, stumps and rubble can form the core and the rest can be built up between the core and a protective wrapping of wattle fencing, log booms or stakes to keep the whole edifice together until it is knitted by plant roots. Turf makes instant topping to hold the above-water surface together. Small islands could be contained by strong wire mesh and stakes. Such islands will be unstable for a while and will need to be monitored. In a shallow pond, the material can be retained in big sewer-pipe rounds and planted with aquatics or island vegetation. To camouflage island cladding, marginals can be planted in raised beds enclosed by pipes, boards or logs.

Rafts

The essential points of a raft are that it can float, but remains where it is wanted. It needs mooring in such a way that it can rise and fall with the water level, yet not spin in dizzying circles in the wind.

Floater materials include logs and planks, old tyres, empty plastic or metal containers, polystyrene and watertight plastic bags like old fertiliser sacks. Of course, any 'empty vessel' will float.

Moor the raft to the bed of the pond, not to the banks (which could give access to rats via the mooring line), and use two anchors at opposite ends to prevent swinging. The anchor could be, say, weighted concrete or bricks, or perhaps an old fertiliser bag full of rocks and stones.

Rafts for waterfowl need to be planted. A wire-netting frame resting just above the water level and suspended by the floats can support the planting medium and allow root access to the water. (It could be fixed permanently on stilts in a scrape pond with a stable water level.) Wattling has the same effect of supporting the vegetation while preventing it from drying out. Hiding places can be created from wire netting, old tea-chests, bits of metal sheeting, oildrums, pipes or whatever else is to hand, camouflaged with straw thatch, reeds, and so on. There are more ideas in Chapter 8.

Linings

Ponds which are not self-confining on impermeable subsoils need to be lined to retain water, unless they are excavated into a high watertable. The choice of practical linings is puddled clay, concrete, preformed fibreglass or flexible sheeting (including polythene, PVC and butyl).

Table 4.1: Pond linings: advantages and drawbacks

Puddled Clay

Traditional material — cheap if local source, bulky to transport.
Quite easily repaired, though time-consuming.
Laborious to apply — requires diligence and time. Labour intensive but no need for special equipment.
Long-lasting if properly consolidated in several layers and if kept from drying out.
Susceptible to damage by cattle hooves unless very thick.
Can be applied to any shape or gradient.

Bentonite Clay

Expensive.
Quick and easy to apply if site thoroughly prepared.
Not good for shallow ponds or steep banks.
Easily damaged by livestock or water-scouring.
Dangerous to fish in its powder form.

Concrete

Requires skill as well as labour, but less labour than clay.
Skill especially important if reinforcement needed (e.g. ponds larger than
8 × 5 × 1m) (25 × 15 × 3ft).
Much bigger hole has to be excavated.
Needs curing before stocking or planting.
Durable if properly laid; not harmed by drought but may be damaged by
frost and ice.
Repairs may be difficult.
Some flexibility of design but sides may be difficult to make.
Should last 50 years if laid well and laid in a day.

Fibreglass

Very small ponds only — limited sizes available.
Expensive.
Very strong.
Difficult to dig hole to fit shape, and total inflexibility.
Can be stocked immediately.

Flexible Liners

Not ideal for very big lakes: difficult to grow fringe vegetation, which
leads to lack of insects and hence less food for other life.
Can be quite easily damaged by trampling boots, livestock, fieldmice
burrowing underneath for a dry nesting site or gnawing and creating
leaks.
Some types prone to deterioration if exposed to ultraviolet light — keep
them covered with water, plants, edge paving, etc.
Problems if watertable rises or land drains flood underneath: the liner
may start to float.
Difficult to handle in windy conditions and can be difficult to seal joins
between sheets or around obstructions such as pipes.
Use black liners to reduce algae growth.
Very adaptable to shape of pond — gives plenty of design flexibility
except for steep banks.
Costs vary — shop around — but cheaper than most other linings, except
local clay.
Polythene is short-lived (2–3 years) and not very elastic.
PVC could last for 5–10 years.
Butyl is said to last for 50 years. Slightly more elastic.

Puddled Clay

Spread a layer of clay (soil with at least 30 per cent clay, as for dams) over the whole site, 7–10cm (3–4in) thick at the most, then moisten it slightly to make it workable. 'Puddle' the layer either by persistent treading with boots, or well-trained sheep, or rollers and tampers. Keep it moist enough to work and to prevent it from cracking. When the first layer is thoroughly consolidated, spread another thin layer and repeat the procedure. Take plenty of time on each layer and do not let the lining dry out between puddling sessions. There should be at least two layers and the minimum final thickness of the lining should be 150mm (6in). Then polish the clay surface with the back of a spade blade. For a small (garden) pond a layer of soot under the clay is said to deter worms from burrowing under it.

Bentonite is an aluminium silicate clay supplied in powder form which swells considerably when water is added and becomes an impermeable gel. It is good for repairing cracks but expensive to apply as a whole liner. The powder is spread on site, raked into the soil, tamped down and protected with more soil.

Gley has been used in some developing countries. Cover the bottom and sides of the pond with pig dung in an even layer and tamp it down. Cover the dung with vegetable matter (banana leaves in the right part of the world, or cut reeds and grasses) and tamp that down. Add a layer of soil, tamp, wait two or three weeks and then fill the pond.

Concrete

Drain the site and keep it dry during the work and the setting of the concrete. The contours need to be gradual and the soil must be consolidated after excavation (scrape out the soil rather than dig it) in order to avoid subsidence, which would crack the concrete. Use a 1:4 mix of Portland cement to sand-and-gravel ballast or, for a larger project, 1 part cement, 2 or 3 parts fine aggregate and 4 or 6 parts coarse (1:2:4 for reinforced work). Mix with water at rates of 25–28 litres of water per 50kg of cement (5.5–6 gallons/cwt).

Small saucer-shaped ponds can be lined with 7.5cm (3in) of cement in one layer. Ponds of, say, 2 × 2m × 15cm (6ft square and 6in deep) need a base thickness of 10cm (4in) with 15cm (6in) on the side walls; a pond 3 × 3m (10 × 10ft) needs at least 15cm (6in) on base and sides.

Tamp the concrete firmly, using a solid board, to compact it and make it watertight. To form the sides, use hardboard shutters and reinforcing rods. Let the concrete dry for at least 24 hours, then fill very gradually with water so that the setting is completed under water. It could be rendered with a mix of sharp sand, cement and a waterproofing powder.

Concrete needs to be 'cured' to get rid of free lime before it can be stocked. The pond should be filled with water, left for a week, emptied and refilled, repeating the cycle several times until the water is clear and scrubbing each time it is emptied. Or the surface can be neutralised with Silglaze, or the pond can be left empty for six months before stocking.

Fibreglass

Dig out a rectangle (which is easier than trying to dig out the shape of the pool), put a little sand at the bottom and set the lining in the hole. Level it with bricks, then gradually fill in the gaps between the lining and the rectangle with soil or sand, ramming it in firmly.

Flexible Linings

These are moulded to the shape of the pond by the weight of the pond's water. The banks should slope at a maximum of 20 degrees to be able to hold the backfill which keeps the liner in place.

Excavate smoothly, trying to avoid creating little pits and humps. Remove any sharp items like stones or rubbish which will pierce the liner. It may be worth spraying on a weedkiller unless a good depth of topsoil has been removed. The site needs to be properly drained (permanently) or the liner might float.

To calculate how much liner is needed, estimate the greatest depth and the greatest length and width. (If the shape is irregular, measure out a rectangle which encloses it.) Allow for a total liner length of twice the depth plus the length, and a total width of twice the depth plus the width.

Cover the base of the pond with fine sand or, for small ponds, thick layers of damp newspaper, peat, ash or sawdust, to cushion the lining. In a large pond or reservoir the sand should be 4–5cm (1.5–2in) deep and the slopes can be cushioned with a layer of polyester mat which is often supplied with the liner.

Lining tends to be supplied in 5–10m (15–30ft) widths and for many ponds the sheets need to be welded together. For large ponds, professional laying and sealing is preferable, especially where there are various outlet pipes and other obstructions where the sealing needs to be very strong. Pipe collars can be used, already welded to the sheeting in the factory. Joins between sheets can be made with mastic or adhesive tape with a 15cm (6in) overlapping of the sections, or by interleaving the edges in a backfilled trench; but professionals use heat sealing.

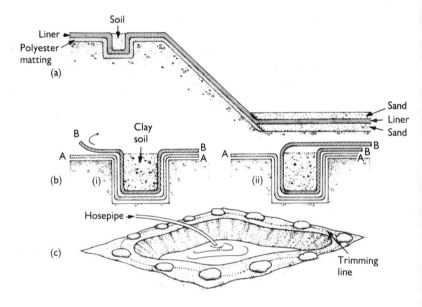

Figure 4.10 **Flexible liners**
(a) *The liner can be held in place in a backfilled trench. Note the protective polyester matting under the liner, and layers of sand sandwiching it on the pond bottom.*
(b) *To join two sheets of liner:*
(i) *Take out a trench and lay the edges of both sheets, one on top of the other, so that they overlap the edge of the trench. Backfill with soil to keep them firmly in place.*
(ii) *Draw the main part of the top sheet up and over through 180 degrees and lay it in its proper position.*
(c) *For a small pond, keep the liner in place temporarily with bricks or stones, adjusting them as necessary while the pond is gradually filled with water. When full, trim off surplus liner, leaving adequate overlap to be anchored and hidden under soil, turf, paving slabs, rocks, etc.*

Great care should be taken not to damage a laid section by treading on it while working on other sections. Protect each piece with a layer of sand as the work progresses.

The linings are designed to stretch a little and are usually placed so that they just touch the bottom, with their outer edges weighted temporarily under stones to keep the material in place while the pond is filled; the stones are shifted as necessary during the filling. Or take out a small trench around the circumference of the pond, just above the final water level; drop the edges of the sheeting into it and hold in place with soil, leaving a little bit of slack to avoid too much stretching.

Fill the pond gently. (With a small pond or a garden pool, let hose-water merely trickle in.) As the weight of the water increases, the lining will be moulded to every contour of the excavation. When filling is complete, the edges can be trimmed off, leaving a surface overlap of perhaps 15–20cm (6–8in), or substantially more for a larger pond, buried under turves or paving. In a garden pond, paving which very slightly protrudes over the pond water gives a shady shelter for fish and wildlife and also softens the outline.

Dabbling ducks or the hooves of drinking livestock will probably expose the edges of the liner in due course, and it may be worth covering them with puddled clay and then building up a bank on top, well turfed to consolidate it. If anglers will be using the pond, put down some protective matting over the edges at specific angling stations to avoid damage by treading.

Spread a good layer of soil over the bottom, partly to protect the liner from ultraviolet degradation, partly to hold it in place and partly to give a planting medium. Put at least 10cm (4in) on the base and twice as much on the slopes. Larger ponds could take up to 30cm (12in) of soil on the base.

Dewponds

Dewponds are small, shallow, man-made ponds up on the porous chalk downlands of southern England, created since at least medieval times to provide drinking water for livestock in grazing areas well out of reach of the watertable. They lie in depressions in the chalk lined with puddled clay (or, later, concrete) and their source of water is largely rainfall, direct into the pond and running off the surrounding area. They are carefully sited for maximum catchment in a cold-spot where the mists naturally gather and are often placed just under a rise (perhaps a natural slope, a bank, a road or a lip) to catch as much shed water as possible, and are embanked for maximum capacity and catchment.

They need maintenance to keep the lining intact and, without such attention, they will probably drain dry so that their average lifespan is less than a century. They can be good habitats for amphibians, especially newts, as long as they are surrounded by rough grass and scrub rather than arable, and as long as they are not allowed to dry out. Unfortunately, due to neglect, they are fast disappearing: there has been a big shift from sheep farming to arable on the downs and the drinking ponds are no longer needed. More than 96 per cent of dewponds known to have been on the South Downs in the early twentieth century had disappeared by the late 1970s, for example, and it is anticipated that there will be none at all by the end of the century, except where naturalists take over their management in the interests of the equally vulnerable great crested newt.

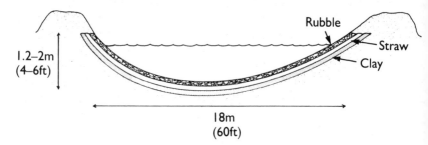

Figure 4.11 **Dewpond construction**
A layer of straw prevents the clay from drying out before the pond is filled. The layer of rubble protects the lining from damage by the hooves of livestock

One of the secrets of a successful dewpond is that its proportions and capacity are such that the rate of evaporation of its stored water never exceeds the rate of collection. To achieve this, the evaporation area (i.e. the exposed water surface) must be considerably less than the rain-collection area. Typical ratios of the two areas are 1:2 or 1:3, depending on the pond's diameter — the higher ratio for the lower diameter. The collection area can be increased by building up a lipped rim around the pond.

Most dewponds are lined with puddled clay, sometimes sandwiched between thick blankets of damp straw to keep the clay moist and prevent it cracking. Protective layers of chalk, lime, rubble, etc. deter earthworm intrusion into the clay and prevent damage to the surface by the feet of livestock. A little treading by sheep could actually help consolidate the clay but cattle soon trample too heavily and destroy the lining.

Fountains and Waterfalls

If water is circulated, by one means or another, it will be constantly replenished with oxygen, to the greater benefit of pondlife. Moving water is attractive in its own right; it creates a dancing of light and its sound can be soothing and pleasant. Anyone who has lived within earshot of a stream or a weir knows how comforting and uplifting that continuous sound of living water can be. It is a great aid to human tranquillity.

In all these senses, waterfalls and fountains can compensate for the lack of a moving stream through a pond. Thelma Seear, founder of the Fountain Society, became passionate about cascading waters and began a crusade to 'serve our heritage of fountains from the past and to develop new fountains'. Her vision is of a world made beautiful with water and though her dream encompasses the grand magnificence of municipal extravagances in the heart of London, or in public parks in Yorkshire and Wales, she also appreciates the 'small, tranquil, perfect, peaceful' fountains that bring to stone cloisters and courtyards their very special atmosphere.

Cascades can be practical as well as aesthetically pleasing and spiritually uplifting. They can be made very simply by putting a log or a rock in the course of a stream to interrupt its flow and let a little head of water build up until it tumbles over the barrier, or they can be marvels of engineering skill involving elaborate designs, powerful pumps and the shifting of tons of beautiful boulders.

The simple cascade is of great practical value in fisheries, both for keeping the water well aerated and for helping to maintain a regular temperature, cooling shallow summer waters and deterring ice formation in the winter. A typical basic system is a series of pools scraped into a natural slope in a descending chain, with spring water dropping from one level to the next over a lowered section of each barrage or through pipes.

Any natural stream, however tiny, can be dammed to create a chain of pools linked by falling water, some of the falls slight, some more dramatic, each pool filling to its brim and then letting the water spill over the edge to feed the next pool, some of the falls as narrow focal points suggestive of gorges, others stretching widely in a shallow, embracing horseshoe curve.

In ornamental garden ponds, fountains and waterfalls really come into their own and give the pond an extra dimension and the liveliness of movement. Not all plants appreciate this; waterlilies, for example, prefer still water, but if the cascade is a trickle rather than a torrent and is carefully sited there will be quiet backwaters for the lilies.

If there is no natural stream or spring, piped water can be used instead and can be circulated by means of a small pump to save extravagant water usage (see Pumps below).

In a garden or park setting a combination of fountains and waterfalls can make a striking setting, with the fountain water dropping in a series of widening circles, each level cascading down to the next.

Water Circulation

The greater part of St James's Park in London is under water and this famous lake supports perhaps 3,000 wildfowl and ornamental water-fowl. The resident population of pinioned ornamental birds is hugely increased by wild birds encouraged by the crowds of Londoners and tourists who take pleasure in feeding them. Ornamental waterfowl have decorated the lake since Charles II established a collection of them in the park.

The birds, however, cause problems. The lake is shallow and concrete-bottomed and one of its major troubles in recent years has been pollution — not from industrial waste, oil or sewage but from the birds and the people who feed them. Such large numbers of birds deposit a substantial amount of manure in the pond, and the bird food (usually bread) thrown into the water by the welcoming public adds to the manure and helps build up a considerable amount of bottom detritus, giving the bacteria and fungi whose role it is to break down organic waste an impossibly onerous task. The birds are threatened by botulism, a bacterial toxin living in the mud, and the coot population is prone to gout from ingesting seagull droppings. The duck population is so high that aquatic plants stand no chance of survival and consequently the water is grossly short of oxygen. In addition to all this, less well-meaning members of the public chuck all sorts of rubbish into the lake and the combination of food, litter, dead tree-leaves and bird manure accumulates in the water, leading to substantial silting as well as deoxygenation. Management of this very public lake became a nightmare: the water was stagnant and looked foul — scummy, murky and smelly.

The lake is fed by water pumped from the artesian wells that abound in the London area but the supply was only used to top up the lake in compensation for water loss by evaporation and seepage. Recently, however, the lake has been considerably improved by the installation of an aeration system which circulates the existing water throughout the lake and gives it a revitalising flow of oxygen. Broadly, the system is a vacuum cleaner which sucks water into its pipes, catching debris in a filter and automatically removing it on a conveyor belt which tips leaves, twigs, litter and so on into a dumper bin to become compost in due course. The filtered water continues through a network of pipes for redistribution into the lake and the effect has been remarkably positive. Fountains help to improve the aeration as well, and also give focal points in the waterscape, and the entire lake has been revitalised. The establishment of plants remains a problem but is progressing slowly: they make plant rafts out of old tyres and earth, covered with wire cages to prevent total destruction by the ducks, who can only reach some of the growing tips beyond the mesh.

The scale of the lake's circulation system is of course large but the idea can be adapted for ponds of any size, and the principles apply to all ponds: keep the level of decaying organic matter low enough to avoid pollution but high enough to supply nutrients for plantlife, protect the plants from the birds, seek to achieve a balanced population of pondlife, and in the meantime help the oxygen levels by circulating water at every level and into every corner and backwater of the pond — and let the fountains and cascades play a part in aerating the water.

Pumps, Rams and Waterpower

Pumps can be driven by hand, by electricity, by diesel or, perhaps more fitting in the context of ponds, by waterpower or windpower.

Garden-pond fountains and water circulation usually rely on small electric pumps which fall into one of two categories: submersible or surface.

Submersible pumps are cheap and simple but not very powerful. The pump is placed in the water of the lowest pond in a series and its capacity to generate a flow is governed by the volume of water in the pond and the height to which the water will be lifted. Installation is simple: the actual electrical connection must be done by a qualified electrician to ensure that all cables and connections are thoroughly waterproof, but otherwise all that is needed is to lay a 1cm (0.5in) hosepipe from the pump up to the highest pool, burying it or covering it with rocks to camouflage it and protect it from frost. For a fountain, the outlet is fixed above water level, then a fountain-head is attached.

Surface pumps are needed for bigger pools: they can give a much greater pressure than submersibles. The pump unit is placed close to the pool and must be housed in a weatherproof container. The electrical connection can be made by a competent amateur, and then it is simply a matter of running one polythene tube between pool and pump and another between pump and waterfall or upper pool.

A small, cheap *aquarium pump* designed to oxygenate water in a fish tank can be used to keep a small pond partly free of ice in winter. House the pump in a convenient shed, run a length of polythene piping to the pond and fix its end about 30cm (1ft) below the water surface so that air bubbles keep a small area clear of ice — just enough to let a couple of ducks have a quick bath.

Pumps in general can all be divided into submersible (deep well) or surface (suction) types. They can also be classified in terms of the mechanism which drives them: they may be constant disposal types or variable disposal types, and they may also be divided between reciprocating, diaphragm and centrifugal types.

Reciprocating pumps may be quite expensive and heavy. They are self-priming and can produce high outputs. They are constant disposal types, delivering more or less the same quantity of water for a wide range of heads. Their power to pump is controlled by the height to which the water needs to be pumped. The ordinary hand pump is a reciprocating piston pump.

Diaphragm or *beam-lift pumps* are typical older pumps, reliable and simple but with only a small output and not always self-priming. They can usually take some sludge, which is useful for emptying ponds.

Centrifugal pumps cope with large amounts of water but most of them are not capable of pumping sludge. They are self-priming and are of the variable disposal type, that is, the pumping rate is altered by changes in the head. For much greater technical detail about all kinds of pumps, refer to Brassington's *Finding Water*.

The simplest waterpowered suction pump is the *siphon*, as long as the water level in the pond is higher than the point at which siphoned water will be disposed of. Take a length of hosepipe or flexible tubing. The aim is to fill the tube with water initially and thereafter let the pressure of the water take over. Either put one end of the tube into the pond and create a vacuum by sucking at the other end until water is drawn right through the tube, or plug the outlet end (a potato comes in useful), pour water in to fill the tube, cover the inlet and put it quickly into the pond water, then remove the outlet plug.

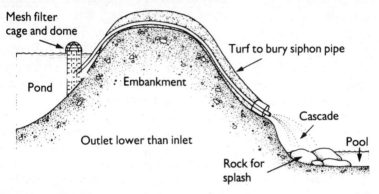

Figure 4.12 *Siphon*
A siphon will carry water out of the pond as long as its outlet end is lower than its inlet end, and as long as there are no air pockets in the pipe. This arrangement shows a siphon used to create a small cascade into a pool: the inlet end is protected from debris, fish and ducklings by a mesh cage and dome; the pipe is buried under turf; the outlet is protected by a piece of drainpipe; and the cascade's fall is broken by a rock to create a pleasant splash and to protect the pool bottom from erosion. If the pond level drops below that of the inlet, air enters the pipe and the siphon will of course stop working. It will be necessary to extract all the air and fill the pipe with water to get the system going again

Wind pumps have long been used by the Dutch to drain wetlands and recently some British farmers have been importing Dutch windmill pumps to lower water levels in the soil. ADAS is investigating the possibilities for land drainage and also for use in wetland conservation projects (contact their divisional office at St Mary's Manor, Beverley, North Humberside). The Centre for Alternative Technology in Wales has built up a considerable expertise on windpowered pumps and also waterpowered machines — turbines, waterwheels and hydraulic rams. Most windmill pumps require a wind speed of at least 10kph (6mph) to operate at all, with optimum operation at windspeeds of 25–40kph

(15–25mph). They can be used to drive reciprocating pumps to store water.

Hydraulic rams for raising water were invented in the eighteenth century. They are valve operated and use the gravitational momentum of flowing water under a small head to lift a small volume of water against a large head, and can be adapted to use a polluted source for power to lift pure spring water without polluting the latter. All a ram needs is impounded water (a pool or catchpit — for example a dammed stream or a pit dug about 1 × 1 × 1.5m (3 × 3 × 5ft) deep) sited above the ram, which is usually a heavy cast-iron or gunmetal engine bolted to a concrete base. A stream flow of about 4.5 litres (1 gallon) a minute is enough to operate a ram. The lift potential is up to 30 times the 'fall' and, given the head, a ram is capable of lifting water to 100m over a distance of 6km (or 300ft over 4 miles). They can operate with a head as low as 0.5m (say 2ft) but are usually worked with a minimum head of 2m (6ft).

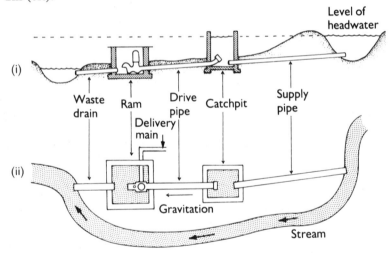

Figure 4.13 **Hydraulic ram**
(i) *Side view of layout;* (ii) *top view*

Rams need to be kept free of debris because foreign matter will interfere with the valves; the device usually incorporates a filter, which needs to be monitored. They need to work continuously and are generally used to push a water supply uphill for storage and use, with a permanent overflow system to maintain the flow when the storage capacity is exceeded. They last for years and years with very little maintenance and are typically used for drinking-water supplies, irrigation, industry, and remote control of sluices and lock-gates.

Water turbines are small pumps driven by streamwater which rotates a turbine that drives the pump to lift water to a reservoir or tank. However, they cannot cope with a higher stream flow than normal, and their numerous working parts need more maintenance than the simpler hydraulic ram. But they are easy to install and are useful for temporary work.

Waterwheels

Those who have ponds have water, and the water can be harnessed for physical work. A very simple application is *water bellows*, which have been used to produce organ music for centuries in Europe and a 1650 publication (*Musurgia Universalia*, by Athanasius Kircher) described how water in vertical pipes drew air down to the organ's wind-chest. A water organ near Salzburg is still blown by bellows driven by a waterwheel today, and in the nineteenth century many church organs relied on mains water supplies to power their bellows by means of a piston-pumping motor.

The *waterwheel* is the type of water machine which most readily comes to mind and it is likely that the Romans first introduced Britain to the principles of the waterwheel (along with rabbits, ferrets, straight roads, central heating and goodness knows what else). Watermills have always been a feature of the British countryside and the typical mill was based on designs by the Roman emperor Vitruvius (first century BC) who invented the undershot wheel. The Romans, and later the Saxons, made their waterwheels of wood, especially elm, the central axle being formed from a whole tree trunk, and it was not until the eighteenth century that cast iron began to displace timber.

Some of our most beautiful ponds are the millponds that were impounded to supply a weight of water to drive corn-grinding stones and iron-making tilt hammers. These controlled water systems had side enterprises as well, like eel trapping and sheep dipping, and there is a whopping cast-iron wheel with a diameter of about 22m (72ft 6in) on the Isle of Man for pumping water out of a lead mine. Most of the old waterwheels are silent now, except for those still working in 'heritage' centres like the Weald and Downland Open Air Museum at Singleton in West Sussex which grinds wheat harvested by horse-power and sells the flour to the many tourists and school parties which visit the museum. Elsewhere, millponds are no longer throbbing with noise and activity; there is no creaking of the big wheel, or crunching of millstones or the swearing of carters and the whinnying of their horses. Instead they are places of great peace, where wildlife can thrive and where humans can contemplate the quiet waters and find tranquillity.

Bridges

Bridges may be purely functional, or can be features that enhance the pond either with subtlety or with the full glory of artistic imagination.

Whatever the bridge, always ensure that it will not impede the flow at times of spate. It could very easily trap branches and debris, creating a temporary dam which floods wide areas for perhaps a considerable distance upstream.

Railway sleepers or discarded telephone poles come into their own for spanning short stretches of water. Their weight makes them stable and self-supporting, and they will have been thoroughly cured in the past so should last a good while and be less likely than other timber to attract slippery algal growth. Plank bridges can become very slippery indeed and may need some kind of footgrip or tread — perhaps wooden battens across the walkway, or lengths of rope or coconut matting. Or give the wood a coat of bituminous paint (of the kind used for asphalt paths) and sprinkle it with fine grit or coarse sand before it dries, and keep off the bridge for at least 12 hours. This treatment will control the growth of algae, which can otherwise be kept in check by adding a cupful of domestic bleach to water in a watering can and sprinkling it on the planks, followed by a good brushing an hour later. There are algaecides on the market, too, but take care that none of these treatments is allowed to wash into the water and cause damage to plants.

Larger bridges need careful designing and building by someone who knows about the stresses involved in spans. Functional bridges can be made of concrete sections on stone or brick pillars, with lengths of galvanised scaffold pipe for handrails: they are not beautiful materials but it is surprising how even these can look attractive in their simplicity if the span is gently arched. The Chinese appreciate the beauty of a semi-circle which, reflected in water, completes itself and becomes the ideal circle that symbolises perfection. The completed circle also symbolises the full moon, and an arching bridge, even without its reflection, creates a crescent, which is a powerful symbol for many different cultures.

Ornamental bridges for water gardens, estate lakes and so on are always more pleasing if they curve, even if only gently. The arch gives a sense of lightness and of airy movement — very different to the horizontal gangplank bridge which looks as if it is waiting for a marching army. Even if the walkway itself is flat, break the outline by using curves in the supporting structure so that the water still reflects an arch.

All kinds of materials can be used — wrought iron, cast iron, wood, plastics, fibreglass, brick and stone. With bridges, the pond owner has a chance to use imagination and flair to create something that is beautiful as well as functional — and that is a balance which applies to so many aspects of pondmaking.

General Pond Management

When a new pond is first filled with water, it will quickly be invaded by algae and the waters will turn murky. That is as it should be and the natural process should be allowed to run its course. Thereafter the basic maintenance of a pond is a matter of common sense tempered by the particular needs of different ponds. Keep silt under control; remove heavy leaf-falls before they rot in shallow ponds; check that trees and shrubs do not outgrow their welcome; keep outflows and emergency overflows clear; keep reeds in check unless the intention is to revert to marshland. Have the water quality checked from time to time and take any action necessary in the light of the test results. Above all, monitor the pond and make a point of carrying out regular inspections.

In conservation ponds, management should be gentle and as unobtrusive as possible. Let the pond sort out its own ecological balance but be ready to interfere if the balance begins to be upset by particularly dominant and vigorous species — rampant reeds, too many ducks, or suffocating algae in an artificially eutrophic pond for example. Avoid emptying the pond completely if it contains living creatures and plants, unless of course it is a fishfarm or watercress bed which have special requirements (see Chapters 6 and 7). Be especially careful to avoid letting a clay-lined pond dry out and crack. Concrete ponds, however, could benefit from occasional draining so that the sides and bottom can be thoroughly scrubbed.

Whatever the lining (or lack of it), keep a careful eye out for seepage at known weak spots and act promptly if there is any suspicion of a leak. Leaks always get bigger unless they are quickly sealed.

A major maintenance routine will probably be the control of algae and aquatic plants, and this subject is dealt with in Chapter 6, which also covers enterprises for growing and harvesting aquatics and water-associated plants and trees, the stocking of plants and anything else to do with pond vegetation. Chapter 7 looks at fish (whether farmed, wild or ornamental) and other pondlife from plankton to reptiles, and Chapter 8 covers birds and mammals. There is a lot more to ponds than water!

Chapter 5
POND RESCUE

Go — pop Sir Thomas again in the pond —
Poor dear! — he'll catch us some more!!
Rev. Richard Harris Barham, **The Ingoldsby**
Legends

It seems that, in Britain at least, the fortunes of ponds in general are on the turn again. For much of this century they have been neglected, abused or deliberately obliterated but there is at last a glimmer of hope for them. People are beginning to appreciate them now that they can no longer be taken for granted, and there is a surge of interest in rescuing forgotten ponds or making new ones. Most of the rescuers are not particularly interested in the practical value of ponds, but are aware that wildlife habitats have been wantonly destroyed and that time is running out for all sorts of aquatic creatures and plants unless active steps are taken to help.

Local papers are full of stories about pond rescue projects covering a wide range of ambitions and costs. For example, in the Buckinghamshire village of Haddenham there is what used to be a famous duckpond, once upon a time the centre of a thriving duck industry. Haddenham's pond became Silly Haddenham's Pond in the nineteenth century when someone built a roof over the water to 'keep the ducklings dry when it rained'. The pond's problems really began about 50 years ago when a local drainage scheme caused the area's watertable to drop (the incidental downfall of many a pond). From time to time they tried to save the shrinking pond but, gradually, the reeds and rushes began to take over and, inevitably, people began dumping rubbish into what was rapidly becoming not much more than an unpleasant swamp. The ducks deserted it and then the villagers woke up to its demise and began a campaign to restore it. They raised several thousand pounds, cleared away the reeds and the rubbish, dug out the silt, relined the bed with new clay, connected the site to the mains water supply, turned on the tap and donated a score of Aylesbury ducks to a local farmer who had grazing rights to the pond. This is a typical village success story: villages are close communities and are adept at generating enthusiasm, funds and voluntary workforces once their imagination is fired.

On a larger scale, the Hampshire market town of Petersfield has a magnificent 9-hectare (22–acre) lake on a heath on its outskirts, a favourite place for walkers and boaters and the site of the annual Taro Fair. In 1986 it was noticed that fish were dying in alarming numbers and it was found that the substantial algae population was deoxygenating the water, which had in any case become badly silted and was only 1.2m (4ft) deep at the centre. In warm weather the fish were virtually cooked. In addition, there was a large wildfowl population, particularly mallards and Canada geese, attracted to the lake by the food which people gladly gave them, and the birds had effectively destroyed much of the aquatic plantlife. The town council was advised to cull and deter the birds (which caused such a storm of local protest that the plan was dropped) and either to oxygenate the water by means of a huge fountain or to dredge the bottom so that the depths would give a cool shelter for the fish and reduce the algae.

For the first time in half a century, it was decided to dredge the lake and to remove fifty thousand cubic metres of silt (about 75,000 tonnes) with the aid of an outsized vacuum cleaner. The bill was estimated at about £100,000 and that was only one of the council's headaches. They were also faced with the problem of where to put all the excavated silt. First they had to clear the surrounding heath so that the stuff could be stored and dried off. Then they spread part of it on the heath as fertiliser (they had been using artificial fertilisers for years in a vain attempt to encourage grass growth on the sandy soil) and offered the rest for sale as prime topsoil to recoup some of the costs. In the meantime the two-year scheme displaced not only local strollers but also the fair which had been held on the heath annually for centuries. Yet such was the amenity value of the pond, and the public affection for it, that the sacrifices were made willingly. The problems, on a lesser scale, are just those which are experienced by many pond restorers.

Within 6km (10 miles) of Petersfield (the area abounds with ponds of all kinds and sizes), three medieval ponds at Waggoners Wells were restored by recruits in the Coldstream Guards, whose combined musclepower made short work of rooting out scrub, clearing back dying trees and building dams and paths with the help of a National Trust conservation group.

In West Sussex, an old millpond at Burton was recently dredged out and the encroaching weeds were cut back in a restoration scheme carried out jointly by the Sussex Trust for Nature Conservation, the British Trust for Conservation Volunteers and the West Sussex County Council — a typical combination of interested groups. The mill itself is an exhibition centre and the whole area is managed as a nature reserve. STNC is well off for old mills and millponds: their headquarters are at Woods Mill, where there is a wealth of waterlife for visitors to see: dragonflies, damselflies, some very tame carp and all sorts of other fish like minnows, stone loach, sticklebacks and other favourites which the children are encouraged to observe in the centre's famous 'dipping pond', which is also home to newts, eels and even a stray American bullfrog. On many of the Trust's reserves, pond restoration is a high priority for their taskforce volunteers.

In contrast to these fairly elaborate and expensive schemes, a typical project is for, say, a Young Farmers group to convert a boggy corner into a pond as part of a farm conservation competition — plenty of hard labour to get rid of superficial rubbish and growth before any creativity can be expressed. At least half of Britain's farm ponds have been lost in the last 50 years and pond conservationists increasingly include farmers who, although they may no longer need ponds for watering their livestock, are none the less voluntary guardians of the countryside, as they have always been. They may start with a simple pond tucked away in a 'useless' corner of the farm but gradually the conservation area is increased, partly to buffer the pond from agricultural interference and partly because one thing naturally leads to another: they create grassy areas for amphibians, copses for birds, and shrub belts as windbreaks, all linked to the pond by hedges and ditches.

The growing interest in conservation ponds on the farm has been encouraged, if not engendered, by the countrywide Farming and Wildlife Advisory Groups (FWAG) which many countries now have. FWAG grew out of the general conservation movement of the 1970s and 1980s and its origins were almost defensive. At the time, it seemed that conservation and agriculture were at war with each other and there was a tendency for conservationists, often born and raised in towns and not truly appreciating the importance of agriculture and the character of those who practised it, to denounce all farmers as despoilers of the landscape and wanton murderers of wildlife. In reality, the majority of farmers love their countryside passionately, if not overtly. Why else would they choose to live and work in it? It is a hard-working life, despite the romantic image of the rustic idyll, and very few can grow rich from it.

FWAG was at first unique among conservation groups in that its members were drawn from both sides of the artificial fence. Wicked farmers rubbed shoulders with students and townbred conservationists, and they found that, after all, they were on the same side. Other groups, by attacking farmers as a body, had made them defensive and that is the worst way of trying to persuade such independent and self-sufficient people to agree with you!

In the late 1980s FWAG is actively and successfully advising farmers who genuinely want to care for wildlife and landscape, and in the last three or four years more and more farmers are interested in conservation ponds, either for their private pleasure or to share with the public as part of a scheme to attract visitors to the countryside with nature trails, farm parks, and so on. Ponds always used to be integral features of farming life and once again they are beginning to find their place on the modern farm. Many are resurrected for practical or productive purposes — fishfarming, angling, wildfowling, irrigation reservoirs and, after the experience of freak weather conditions which disrupt piped water supplies and electricity, even for emergency livestock drinking water and as a source of waterpowered generators. Many more, however, are simply conservation ponds in damp corners which in all probability were originally farmponds anyway.

Farmers of old also made full use of village and roadside ponds for watering their cattle on the way to market, or for watering horses and soaking cartwheels, and these ponds, too, are being given new leases of life all over the country, especially since the impetus of the Save the Village Pond campaign sponsored by the British Waterfowl Association and the Ford Motor Company in 1974, which shamed many a community into clearing the rubbish out of the village pond and creating a feature out of an eyesore. At that time the author of the campaign's practical handbook estimated from Ordnance Survey maps that there must have been about 338,000 ponds in England and Wales; since then no doubt a large proportion have disappeared, though quite a few will have been saved and restored.

Perhaps the most practical organisation of all in the rescue of ponds (and many other habitats and landscape features) is the British Trust for Conservation Volunteers. The Trust brings together gangs of people actually prepared to get out and *work* on conservation projects rather than simply stand dryshod on soap boxes. They also publish a wealth of down-to-earth advice on all aspects of conservation, and their publications are essential reading for any conservation project. In the context of ponds, the bible is BCTV's *Waterways and Wetlands*.

These and other organisations listed in Appendix E are important sources of advice for those who want to renovate or create a pond and some might also be a source of grants if a project merits backing. Many county councils, for example, want to encourage farmers and landowners to make ponds; parish councils might well be inspired to give grants if enough villagers volunteer to clean up the village pond; large conservation-conscious companies are always worth approaching, particularly oil companies, chemical industries and car manufacturers and others with slight consciences that their profits might be gained at the expense of the environment.

But first, of course, find a pond to renovate. If the SVPC figures are fair, there must once have been at least a couple of ponds within nearly every 2.5 sq km (square mile) of lowland Britain. Many have disappeared for ever under concrete and tarmac but many more have simply silted up over the years and regressed to swamps, willow carrs or damp patches in fields and gardens that never seem to drain properly. Obviously a damp area set in a slight basin could once have been a pond.

For a start, take a look at old maps. In fact, because the victory of pipe over pond has been so recent, many currently available Ordnance Survey maps still show ponds that have long since disappeared, especially in areas where no completely new survey has been carried out since 1936. Use the large-scale 'parish' or 'farmer's' maps; the 1:25,000 series (originally 25in to the mile) covers an area of 1,000 sq m and shows everything except contour lines, and marks every enclosure with a note of its area in hectares, including ponds. Contours are shown on the 1:10,000 series (the old 6in maps).

Look along streams for possible sites; watch out for the more obvious places like old mills, hammer ponds, monastic sites, farmyards, areas of clay where there used to be brickmaking and so on. Look for names connected with iron-making (hammer, forge, furnace, cinder). Check local history sources for references and in particular visit the county records office to look at its collection of large-scale estate maps which often used to mark even every tree, almost every bush, and certainly every ponds; some of the maps may be 300 years old or more.

If a likely pond is already known, find out who owns the site. If it is an old village pond, for example, it might be owned by the local council (talk to the parish clerk), or by the brewery if it is next to a pub, or by a farmer or land-owner, or perhaps a nominal (and probably absentee) lord of the manor.

Before proceeding with any pond renovation project, it is essential to carry out a careful survey of the site and the factors that affect it. Even before that, very careful consideration should be given to the aims of the project. The first question is 'Why bother?' and the next is: 'What for?' A pond dedicated to the conservation of wildlife is not the same as one which is primarily devoted to a particular productive purpose, though in many cases productivity and conservation can go hand in hand.

Decide what has already gone wrong with the pond, and why. Look for likely problems to see if renovation is feasible or if the site is best as wetland rather than pond. Work out what the pondlife you would like to see really needs, and come to a decision on priorities. A 'conservation pond' is not necessarily all things to all pondlife. For example, if the scheme is the reclamation of a village pond, there may be a conflict of interests between the villagers who want to visit the pond and the wildlife which would prefer to be left in peace. If you are trying to offer a home for amphibians, do not have any fish in the pond because they will eat most of the tadpoles. So might the ducks — but forget the old wives' tale about ducks eating adult goldfish and trout. If you do want to encourage plenty of ducks, you must keep them well out of the way until the pond's new plantlife is thoroughly established because if there are too many ducks the plants will not stand a chance, and without plants a pond dies.

Above all, think before you act, and have very clear aims.

Conservation Ponds

First of all, look at the pond in the context of a locality. For example, if it is in farmland look at the farm to see where a conservation pond fits into the whole scheme of things. Are there other ponds (or potential ponds), watercourses and wet areas on the farm? If so, can the separate aquatic environments be linked so that wildlife can migrate as it wishes, and can a variety of such habitats be created by careful planning and management — perhaps an open pond here, a sheltered one there, a woodland one elsewhere, and so on? Variety is the essence of the traditional countryside and a variety of habitats will encourage a variety of wildlife, if that is what is wanted. In special situations, where a rare species exists, the scheme should be designed in the interests of that species first and foremost.

Pond Sampling

Equipment

Pond net (simply made, using fine-mesh nylon curtain material on strong wire frame with long, strong wooden handle).
Mud net (e.g. domestic hand-sieve).
White tray for specimen identification, deep enough to take 2.5cm (1in) or so of clean water.
Glass-bottomed box for observation in water.

1. Take a visual check of marginal life: plants, insects/nymphs on marginal plants, etc.
2. Take a visual check of surface life: plant species, surface insects. Use glass-bottomed box to look into the water: lie on your stomach on the bank.
3. Make sweeps with the pond net from the banks, skimming the top layer of the water for insects without disturbing any mud. Transfer catch to tray for identificaion.
4. Make sweeps in the water among plants.
5. Make sweeps into the top of the silt, gently turning over stones, etc. to dislodge hiding creatures and sweeping the net towards you. Replace stones in original position.
6. Use the mud sieve to scoop a little deeper into the silt.
7. Release specimens as soon as possible after identification, placing them back where you found them.
8. Repeat the process in several parts of the pond.

For a more detailed survey, obtain/borrow equipment for measuring pH of water (colorimetric solutions or papers), water temperature at various levels, water flow (if any) at various points, dissolved oxygen content of water. Take readings in several different areas, and likewise take bottled samples of water from different parts for laboratory analysis if required.

Assess the actual site. See what wildlife already exists and take special note of interesting species (any change in habitat caused by renovation might prove harmful to them, and just the stress of human disturbance may be enough to destroy their sanctuary). Carry out the survey over a whole year so that seasonal changes can be noted, and spend a lot of time looking at the site during May and June, when wildlife is most active and obvious, but take care not to disrupt breeding creatures.

Lie flat on your stomach on the bank and look long and hard into the water: really get down to pond level. Remember that much of pondlife is very small, or well hidden, and remember too that what looks like just another piece of pondweed floating by could in fact be a rare species of plant, or that a small aquatic creature might be the nymph of a rare species of dragonfly. Life is at its most diverse on the waters' edges of the world, where the aquatic mingle with the terrestrial, and a pond, so much smaller than seas and lakes, has a high proportion of water's edge in comparison to its volume and area.

If necessary, ask a local naturalist to help with the species survey or contact local field centre, nature trust or appropriate national organisations like the Royal Society for the Protecion of Birds, the British Waterfowl Assocation, the British Herpetological Society and others listed in Appendix E. Then decide which of the existing wildlife and natural features need to be encouraged or improved. If rare species are found, contact the local nature conservation trust or the Nature Conservancy Council for expert advice. The area may already be a Site of Special Scientific Interest (SSSI), in which case it is legally protected and permission must be sought before any work of any kind can be carried out. If there are any protected species, it is an offence to disturb, kill, injure or remove them. (You may not even so much as take photographs of specially protected birds at or near their nesting sites unless you have a licence from the NCC.)

Look at the pond's surroundings to see if there are perhaps too many trees or too little cover and access to other habitats. A pond must be open to the south for maximum sunlight for photosynthesis in the water, but with a few scattered shaded areas provided by occasional trees and shrubs on other banks. The branches will also provide good niches for insects, which is good news for the fish that wait for them to drop into the water. Fish will also appreciate access to shade in hot weather, but any bankside trees should be carefully managed by coppicing or pollarding to keep their growth in check and avoid too much shade over a small pond or too much decaying leaf-matter in the water. Ponds should be sheltered, however: leave a shelter-belt of trees well back from the water to take the worst of the wind, and plant a copse of berry-bearing shrubs between the belt and the pond to give food and cover to birds and insects.

Be aware that attempting to alter one aspect (removing vegetation, altering banks, adjusting depths or inflows, etc.) could have ramifications on other aspects. Like playing chess, playing with ponds needs to be thought right through in advance.

Use proper surveying equipment to plot the contours of any proposed excavation and ensure that renovation does not in some way cause unexpected flooding — at the pond or elsewhere in the area. Beware of over-enthusiasm: in most cases leave the pond's existing outline and contours as they are but plan to have gently sloping banks on at least one stretch, to encourage a maximum variety of marginal plants and also to give young amphibians, ducklings and undignified ducks easy access to and from the water.

Check the water: its source, its quality, its plantlife and animals, its possible pollution and the stability of its levels over a year. Ask the water board, environmental health officer or FWAG adviser to help with water quality tests if you are in doubt. Watch out for obvious pollution — dead fish (toxic substances, disease or deoxygenation), smells, foam or oily slicks — and signs of remediable eutrophy (excessive growth of algae and duckweed, invasive rushes and reeds, smelly gases from an excess of decaying matter on the bottom, and so on).

Look at the local geology and topography to see what kind of soils and substrates there are, what sort of natural drainage due to the lie of the land, how big the catchment area is and what sources of pollution it might contain. Will the pond be subject to drought or to flash-flooding and, if so, can it be protected from these extremes?

Check the existing pond bottom, estimating the depth and content of the silt and the type of base material. In a really old pond, it might even be possible to deduce something of its history by taking core samples of the mud and examining the different layers.

The pond's shape (and its name, if any) may give clues about its past. If it seems rather regular, it was made for a purpose. Perhaps it is one of a series of fishponds, or rectangular watercress beds, or a circular dewpond or part of an old mill system. Steep sides imply either a purpose-dug pond not intended for watering livestock, or that the pond was in fact a pit where material of some kind was excavated (clay for daub or bricks, marl for the fields, etc.) and it became a pond incidentally. If there are no embankments, then what happened to all that spoil? Consider why the pond existed on the site in the first place and then consider why it has deteriorated to its present state: there may be very good reasons for its demise and it may be quite impractical to try and renovate it.

A pond's history is of more than idle interest. It gives clues as to what the pond can best be used for, quite apart from possibly triggering off an interest in environmental archaeology and local history.

The next stage is to pause for thought and draw all the findings together for consideration. Take a look at the Pond Survey Checklist, which will give you a framework.

Table 5.1 Pond survey checklist

Origins

Why, how and when created?
Subsequent history and use?
Archaeological interest?

Site Assessment

General environments (farmland, village, near a town etc.).
Immediate surroundings and links with other habitats.
Soil type, natural drainage and contours in catchment areas.
Water source: present or potential — location, type, reliability of supply, influences (e.g. upstream users, extractors affecting watertable).
Water quality: pH, hardness, degree of pollution, sources of potential pollution.
Present pond profile:
   Type and condition of bottom, degree of silting/rubbish.
   Shape and contours — shores, depths, islands.
   Size — area and depth of water.
   Plant species present (varieties and abundance).
   Animal species present (varieties and abundance).
      [In both cases, within and around pond.]
   Access of light to water surface.

Ideas for Site

Main purpose.
Additional possibilities.
Extended schemes (e.g. part of a nature reserve/farm trail; one of a series of fishponds, etc.)

Scheme for Site

Present problems and solutions.
Priorities.
Sources of advice.
Sources of funds.
Sources of practical assistance.
Consultations with authorities (planning permission, water extraction, impounding licences, fish movement, etc.)
Detailed plans: profiles, work needed, order of work.

It may be that the pond in its existing stage of regression is in fact valuable to the wildlife which finds sanctuary there. Wetlands are important habitats and, although a swampy area may not be as aesthetically pleasing as a cleared pond, it does lend variety to the landscape and to wildlife. It may still need management to check further regression but drastic dredging and clearance could do far more harm than good, especially if there are no other swampy patches in the area. Ponds should never be considered in isolation but as part of a broader environment.

Any renovation should be in stages, to minimise damage to wildlife. For example, there have been cases of important dragonfly ponds losing virtually their entire population of these beautiful creatures because either the weed clearance was too sudden, or drainage for a long period to allow dredging dried out the habitat, or the insects' hawking rides in surrounding woodland were destroyed. Time any work to avoid breeding seasons or vulnerable stages of plant lifecycles and carry it out little by little, discreetly and gently.

What, then, is the aim of the renovation? Is it to create and maintain a private nature haven for any lifeforms which care to make use of it? Or are there particular species which already use the site and are rare, locally or more widely, and in need of protection? Is it to be an 'amenity' pond to make part of a village look more picturesque? Is the pond part of a much wider scheme? If it is being considered in isolation, that is a big mistake: there are many external factors which will affect a pond and, equally, the pond itself can affect downstream environments or alter the ecological balance of an area. There is a lot more to a pond than water for some pretty ducks.

Draw up some kind of a management plan so that the aims are clarified; make an attempt to cost the project and then consult bodies that are reliable sources of advice on pond restoration, or those that might give grant aid — and talk to them *before* any work is put in hand.

Water Supply

For the sake of pondlife, *stability* of water supply is important, both for maintenance of water levels and for quality. During preliminary surveys it is essential to test the reliability of the source; a pond which dries out every summer is no home for wildlife and this may be why the pond is derelict. If water supplies do fluctuate to extreme it is possible, but expensive, to devise drainage or pumping systems that ensure a constant water supply (or even make use of a mains supply if it is not too heavily chlorinated) but it may be wiser to treat such a site as wetland rather than a pond. Where conservation of wildlife is the aim a waterlogged swamp, marsh, bog, fenland or flooded meadowland may be more valuable than open water.

Water may be disappearing because of problems with the pond itself, and these are more easily solved. Perhaps there is a leak in a dam or embankment, or perhaps the pond bottom has been breached. The feeding spring may have become clogged with silt, or have diverted itself elsewhere — for example, down a swallow-hole.

Clearing the Surroundings

Pond renovation can seem daunting when the prospect is a tangled mass of scrub, weed, fallen branches and rubbish all mingled in with fairly smelly ooze. Think before wading in with a chainsaw and slasher, and proceed with caution. Beware of the urge to be tidy and make a clean sweep. Clear gradually, preferably in several stages over a long period to minimise widespread damage to existing wildlife. Avoid the breeding season for such work; late autumn and early winter are better anyway because deciduous leaves will have fallen so that the 'bones' of the site can be seen, which is much less daunting than fully clothed scrub and less cumbersome to deal with. It helps to mark which trees are to be cut and which removed during summer when the full effect of their shade and area of leaf-fall can be seen and their species more readily identified.

Conifers should be removed: their needles tend to make pondwater too acid (pH 5–6). Leave a few alders or willows to provide summer shade which will give fish a cool lie among the roots and will also help to keep some areas clear of algae and duckweed. Coppice or pollard the trees, especially those on the banks which might provide too much shade over the water and might also be so heavy in the crown that wind-drag begins to loosen their roots, destabilising the banks and eventually bringing the tree crashing down into the water. Incidentally, moorhens appreciate a fallen tree lying half submerged: its bushy crown makes a good nesting site for them, and ducks often loiter in the shelter of the branches, too. An old tree trunk, half submerged, becomes a popular loafing log for many waterbirds and even the youngest can clamber up its slope out of the water.

If there are too many trees near a pond, air movement will be suppressed and this will not help oxygen supplies in the water. A lot of trees means a lot of leaves, too, and they could easily overload the pond. Willow leaves in quantity can become toxic to fish: they contain the substance from which aspirin is made. Trees on dams should not be allowed under any circumstances as they will weaken the structure and their dying roots will create water channels that will gradually drain the pond and ruin the dam.

Trees can cause unexpected problems. For example, the devastating gales in October 1987 brought a large pine and an old oak tree crashing down into a big pond containing carp. The branches gouged into several years' accumulation of silt, suddenly releasing a large quantity of methane gas from the bottom debris, which killed off many of the carp. Others died from the release of toxins in the pine resin.

Removing whole, living trees is a skilled task best left to the experts, especially if the stump is to be uprooted (which will leave an ugly hole and it may be necessary to use earth-moving equipment in due course to level a grubbed area). Stumps can be left to rot down very slowly, or the process can be hastened with proprietary stump treatments. Species like willow will, of course, sprout vigorously from the stump unless it is treated, and several other trees react in the same way.

Go gently with the slasher when tackling undergrowth and beware of the urge to make a clean sweep so that everything looks 'tidy': that is only appropriate in formal settings. Wildlife needs the hiding places, breeding places and sources of food provided by patches of tangled vegetation, dead branches, rotting stumps and living shrubberies. Many birds feed on grubs living in decaying matter; amphibians and dragonflies need cover near the water. Dragonflies like a reedbed which gets full sunshine but is sheltered behind by woodland bisected by hawking lanes.

A tidy landscape will be a barren one, and sudden exposure could alter the way water runs off the surface or seeps down to the watertable. The net result could be heavy silting, as anyone who has ever had a pond downstream of major forestry clearance will know.

One way of clearing undergrowth fairly comprehensively in control-led patches is by putting in a few pigs. They do an excellent job on brambles, nettles, bracken and so on and will leave the ground well cleaned and the soil well dug and dunged. Their access can be controlled with a single line of electric fencing, with a swathe cut through the undergrowth to prevent the current shorting through vegetation in contact with the wire. Bear in mind that pigs *can* swim, and certainly love a good wallow in a boggy pond! Choose a long-snouted type, like an uncontrollable Tamworth, and make sure that they are *not* nose-ringed (ringing is supposed to deter them from rooting). Use them sensibly, clearing small areas at a time and leaving plenty of growth intact between the patches: pigs are very thorough. Of course, if the pond is to be put to some practical use it may be necessary to clean the surrounds entirely, and then the pigs can be given their heads. They may take longer than heavy plant but they'll be much cheaper to run and could give a good side of bacon at the end of it all. For conservation purposes, however, avoid extremes and leave plenty of scrub; clear the pond margins if necessary, and make a few access routes — for amphibians and birds rather than for people. Indeed, people should be kept away from conservation ponds as much as possible for the sake of the wildlife, which is put under stress by disturbance. In the design stages, decide whether this is to be a true conservation pond dedicated to the wildlife which use it, or is to be peek pond for naturalists, a dipping pond for children or open to intrusion by all and sundry and never mind the wild inhabitants.

Clearing the Pond

Where trees have invaded the pond itself, the problems are greater. Roots penetrate the pond bottom and are potential areas for water leakage, even when living. When they die back, they gradually rot and create channels which effectively pull the plug out of the pond.

It is tempting to leave trees as features, or at least their stumps as miniature islands for duck-nesting, but one day they *will* cause problems. The pond bottom is vital: it must remain watertight unless the pond is well into the watertable or there is a powerful perpetual spring to make good any loss from seepage. To root out such trees is a major operation, because the bottom will have to be made good again and thoroughly consolidated, but it might be better in the long run to face the task now rather than hope for the best later. If watertightness is not a concern, a tree can be left where it is (unless it gives too much shade) and can be turned into an island by surrounding it with earth-bags or retaining logs backfilled with spoil and soil, but only if the island's surface is at the original ground level or else the tree will probably die.

A more likely problem is that the pond is full of rubbish or choked with aquatic plants, and in both cases hard labour is necessary. Be careful when taking out rubbish: there are obvious dangers to handlers if there are sharp, rusty bits of metal lurking in the depths, and there is always a possibility that containers of toxic matter are in there, too. Old tin cans, whatever their original contents, may themselves be toxic to pondlife because of the leaching out of metal ions into the water, killing off animal life and deterring visiting waterfowl so that the plantlife grows away unchecked and chokes the pond to death.

In the case of aquatic vegetation, do not go for complete eradication. Avoid using any kind of herbicides: they are too destructive, they may have long-term effects which are not yet appreciated and they could easily have downstream effects. They could also well lead to deoxygenation of the water because the plants are killed suddenly *en masse* and, if lef to rot, will make excessive demands on the water's dissolved oxygen while at the same time the means of restoring the oxygen through photosynthesis have been summarily removed. The restored pond will be very much in need of its oxygenating plantlife.

Chapter 6 gives details of plant removal and control. Free-floating plants like duckweed and algae can be raked or dragged off the water; they are persistent and if there are nutrients and sunlight they will soon bloom again, but at this stage clearance is really for the sake of being able to see what is in the pond. Rooted plants can be cut or dug; cutting encourages subsequent regrowth, if this is wanted. Before undertaking extensive digging out (unless it can be done selectively), carefully lift out any plants to be saved, with their roots encased in pond soil, and put them in a shady corner covered with moist leaf litter until they can be replanted.

Remove as much of the raked, cut or dug material as possible, because if vegetation is left to rot in the pond it will have the decomposer bacteria working overtime, depleting the oxygen and building up a layer of muck on the bottom. Leave the material on a bank for a short while so that any stranded aquatic creatures can find their way back into the water. Let it drain anyway so that it is less heavy to handle. Cut or raked material and silt make good compost: heap in the usual way and leave it to rot down before using as a mulch.

Dredging and Shaping

Dredging might be the answer to an ailing, ageing pond's problems. It gives extra depth which tends to allow areas of open water to develop; it removes accumulated debris and rubbish in the silt which may be a factor in pollution by creating anaerobic sludge on the bottom; it delays or reverses regression to reed-swamp in eutrophic ponds; it is a fairly drastic way of removing a lot of unwanted rooted plants all in one go. A well-managed pond is dredged to some extent on perhaps a ten-year cycle, working gradually across the pond in that period rather than doing the whole thing at once, so that existing wildlife has a chance to redistribute itself. Although there seems to be less life in a pond during the winter, remember that many creatures and plants may be hibernating in the mud.

Any decision to dredge should be taken after much careful deliberation. Dredging is a major disturbance to existing pondlife. For example, some insect larvae are aquatic for as much as two or three years before they change into terrestrial or semi-aquatic adults, and damselflies a couple of years hence. Removal of shallows may destroy perfect breeding habitats for frogs which have come to the pond for generations.

In the planning stages, think about what to do with the spoil — there will be a *lot* of it. (Try digging out a post-hole: the amount of soil removed always looks much more in the barrow than one can conceive from the size of the hole excavated.) Dredging can be done by hand, if the pond is shallow and the silt not so deep that it pulls your boots off, but it is very hard work. There's the actual digging and then there's the disposal of the spoil, either by throwing up each spadeful onto the bank (space soon runs out) or using a team of people with buckets or barrows to ferry the heavy, wet stuff away, preferably without causing too much damage to the pond surrounds.

In some cases the job is much quicker, easier and more effectively done with mechanical aid, with the added advantage that the period of disruption to wildlife should be short, though noisy. Great care must be taken to minimise that disturbance and also to preserve the bottom lining intact, particularly if it is puddled clay.

Do not drain the pond dry before dredging, but naturally it will help if the water level is dropped to some extent before the work begins. This might be difficult: unless the pond already has a workable sluice or plug, the water will probably have to be siphoned or pumped out and then there may be a disposal problem.

Mechanical equipment can usually be hired with or without an operator. There are small, light models which are useful where access is difficult. Make sure the operator understands exactly what is required by way of depth and shaping and be prepared to put down protective tracks to prevent damage to the pond's surroundings. If shaping is part of the work, survey the site carefully in the planning stages and stake out the outline.

Shaping should include any adjustments to the banks, creating shelving for different planting levels. Gentle shelving has several advantages: it allows a wider variety of marginal plants to become established and it generally gives a more secure bank. It also allows ducks to dabble and gives access in and out of the water for short-legged little creatures like ducklings and amphibians. Unless instructed otherwise, many excavators will make nice, tidy, vertical banks so the matter should be discussed with the driver beforehand.

The same considerations about sloping apply to existing islands, which are favourable nesting sites. Care should be taken to ensure that the young broods have easy access (5cm (2in) is too steep a cliff for a duckling — it really needs a very gentle ramp right into the water) and also to ensure that the island remains beyond reach of most predators. A distance of perhaps 25–30m (80–100ft) from the shore should deter mustelids (except mink) and foxes, though the latter are quite capable of swimming. Landscape the island so that there is a south-facing loafing bank for sun-bathing, with a low wind-break set against the prevailing wind.

Excavated spoil is not ideal for creating islands (see Chapter 4) but it could be distributed judiciously on existing islands above the waterline — again, little by little, as islands have their own flora and fauna. Spoil can be rich in nutrients and it could be spread on the land to good effect once it has dried out a little and as long as it is not full of rubbish. Do not simply pile it up on or near the banks (or those of other watercourses): it will wash straight back in again with the first heavy rains and in any event will probably take quite some time before it can be planted or grassed. It might be best to tip the material elsewhere: talk to the local authority.

In many cases it may be that excavatory dredging is unnecessary and all that is needed is removal of the type of black sludge which accumulates when large quantities of leaves decay in a pond over a long period. Then the answer could be to pump the muck out with a hired sludge gulper, and use the sludge as manure on the land.

Bottom Repairs

If a pond has a tendency to dry out, the watertable may be too low or the pond's lining or banks have been breached. To trace a leak, sprinkle a powdered dye on the water surface near the suspected breach. The escaping water will draw the tracer towards the leak, often quite dramatically pointing the way. It helps to put a white tray on the bottom so that the dye is more clearly visible in murky water (let any stirred up silt settle again before adding the dye). People used to use potassium permanganate as a tracer — its bright magenta is a very clear marker — but it should be treated with caution because in any quantity it is a disinfectant (it is used for water purification) which could kill aquatic life; indeed the water authority would strongly advise against its use. The recommended alternative is fluorescein, a vegetable dye, or, if nothing else is handy, a little brightly coloured ink. If the path of the dye on the water does not obviously pinpoint the leak, drain the pond and look for the dye's colour on the bottom, where it will have traced any cracks.

A cracked concrete pond can be repaired with a liquid sealant over the entire surface (naturally all plants, fish and other life have to be removed first and the pond drained), then soil is added once the coating has dried so that the pond can be replanted. An alternative is to use puddled clay or a proprietary repair product on the cracks, or to lay down a new butyl liner over the entire pond.

Cracks in clay ponds need repuddling, which is skilled work. Cracks may occur if the bottom has dried out at some stage and you should beware of keeping water levels in clay ponds so low that the puddled bottom is exposed at the margins, or of allowing cattle to trample it.

Use greasy, sticky puddling clay, which forms a tight, moist adhesive lining when worked. It needs to be moistened and thoroughly beaten into the crack and overlapping it by a good margin, and finally polished with a spade to make it really impermeable. (See Chapter 4 for details of puddling.) Pay particular attention to areas where ducks have dabbled at the clay, especially where the bottom meets a dam or makes an angle with a bank.

Banks and Dams

The word 'pond' originally implied impoundment or enclosure of water. Many ponds have been created by damming or by building raised embankments, and these (especially dams) can be subject to considerable stress. If they leak, the pond level drops. What begins as a hairline crack is gradually eroded by seeping water until there is a major breach, and if there is a very sudden breakthrough the results can be catastrophic both for the pond and downstream. Dams and banks must therefore be thoroughly checked during renovation and repaired where necessary.

Leaks can be traced with dyes. A typical weak point is where the dam meets the pond bottom and the pond must be lowered beyond this level so that the dam can be inspected thoroughly.

Another common weak point is at the edges of any weirs, sluice gates, outflow pipes and drainage pipes. At all these points water is channelled and therefore applies pressure (more by its bulk than by its current). Check that there is in fact an overflow system for times of flood, to relieve pressure on the dam, and make sure that it is always kept unblocked.

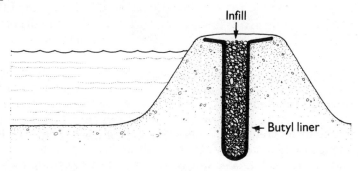

Figure 5.1 **Repairing a breached embankment**
A method of repairing a breach in a small embankment: dig a vertical trench to a depth below that of the pond bottom; line it with butyl and backfill with soil: or make a concrete core

With a small pond, a seeping dam can be partially waterproofed by digging a narrow core trench along its length and through its heart to below the level of the pond's bottom. Line the trench with butyl sheeting and backfill it, preferably with a clayey soil, or plug it with concrete. If an earth dam has actually been breached, however, its repair needs the advice and supervision of an experienced engineer.

Dams are automatic silt traps but unfortunately they build up the silt where it is least wanted: in the pond. If silting appears to be a problem, make a silt trap outside the pond at the inflow. A broad reedbelt acts as a natural silt trap.

Banks often need some attention during renovation work on old ponds: they may have been eroded by livestock or too many waterfowl, or tunnelled by water voles and water shrews. In a conservation pond, banks do not need to be uniform and tidy and indeed natural slippages on steeper banks can form new breeding ledges for insects. To encourage bank-nesting birds like martins and kingfishers in sandy areas, make sure one bank of the pond is steeply graded. Artificial nesting tunnels made of drainpipes can be built into the bank if the site is not alrady in full use.

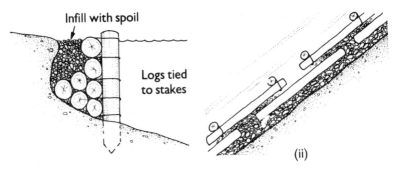

Figure 5.2 Bank repairs
Eroded bank made good by infilling behind logs tied to stakes. The top of the infill should be planted for improved stability.
(i) Side view; (ii) top view

To reinforce crumbling banks, there is a choice of materials according to local availability. Timber is the most obvious, and in water the most useful and durable woods are elm (lasts for centuries, if you can find any), oak (think of old sailing ships) and alder, a natural waterside tree. Do not creosote the wood: it will pollute the water. If stakes are used, beware of driving them through any pond lining. Fill in behind the stakes with wattle fencing (weave wands of hazel or willow cut from the pond surrounds), branches, poles, brushwood (temporary), reed 'mattresses', old railway sleepers, scaffolding boards, etc., all backed with packed-in spoil, sacks of clay or concrete, sand bags, corrugated iron, and so on. If stone is readily available, build a drystone wall leaning into the bank.

Grassy banks should be left for waterfowl sunbathing, preening, resting and grazing, and it may be worth providing a sandy or gravelly slipway where they leave the water to avoid the grass being paddled to a muddy slope by the frequent march of wet webbed feet. A few sheep (especially primitive breeds like the Soay, if they are willing to stay around — they have minds of their own) or Chinese geese will keep grass on banks and dams under control, and sheep sometimes also graze reed shoots, but the animals may have foot trouble if they stay long on wet land.

Some wild rice (*Zizania aquatica*, an annual imported from Canada) could be planted to provide cover and shelter in marshy margins until the natural vegetation can take over. Waterfowl love its grain, and so do fish.

Colonisation and Stocking

A renovated pond will stock itself quickly as long as clearance and dredging have not been too drastic. Even a dried-out pond probably still harbours viable seed which will germinate once the water is replenished. There must be adequate vegetation before animal life can thrive. A pond newly filled with water will soon ʾecome a thick, green soup and that is *good*! Leave it, however unsightly it may be for a while — especially during its first summer. It is an algae soup, the first link in the food chain, and it will clear of its own accord once the pond has established its own ecological balance.

The animal species will arrive once the pond has settled down. Some fly in, some crawl, some drift in on the current from the feeder stream, some hitch a ride on the legs of incoming waterfowl. If there is a healthy pond in the area, a few pailfuls of water and silt will introduce small animal life more quickly to the renovated pond but permission must first be sought from the owner.

There are very strict regulations about transferring fish from one piece of water to another (see Chapter 3) or removing specimens of any kind from a nature reserve. If you want to transplant some frog spawn or toad spawn, make sure that you are not raiding the only pond for miles around which has spawn in it and thus depriving it. There are often advertisements in local papers or pet shops offering free frog spawn from garden ponds.

Some creatures do not cohabit! Fish, for example, devour tadpoles in huge numbers and also make substantial inroads on the planktonic animals that are the bread and butter of so many pond inhabitants.

Don't introduce fish until other life has had a chance to get properly established, both for the sake of that other life and also because the fish will soon go hungry.

Seeds and fragments of self-rooting weed will be carried into the pond unwittingly by ducks or fish, or will drift in from the feeder stream. If you are collecting plant material from another pond, do not be greedy: take a little here, a little there, to avoid upsetting the balance in the donor pond. If someone somewhere is dredging out a ditch, pond or river, there will probably be plenty of plants going begging and you could gather up a few bucketfuls of rhizomes and silt.

Whatever you decide to import in the way of plant or animal life, make sure first of all that the recipient pond can offer habitats suitable for the species.

Managing a Conservation Pond

Once a pond has been renovated, the aims of future management must be to keep an eye on the balance of pondlife and to watch out for adverse external influences like pollution or watertable interference. The main problems will probably be the control of pond vegetation, bank vegetation, silting and pests.

In a conservation pond, however, the intention should be the least possible interference, within reason. A gentle dredging cycle is sensible and plant control needs at least an annual survey. Algae explosions need constant observation and quick action where it is really necessary, but bear in mind that algae are a vital and basic part of the food chain in a pond: they are decidedly an asset unless nutrient levels are artificially and suddenly increased.

A gentle programme of weed control could save a major clearance operation later on. Cull the vegetation judiciously, in small areas and infrequently to minimise disruption to the pond's inhabitants. If a species is becoming dominant to the detriment of other plant varieties, be more ruthless. Keep an eye on the relentless reed invasion (someone has probably written an ode beginning 'Oh Phragmites!') and take steps to deepen the adjacent water to at least a metre (3ft) as a barrier to its progress. There are ways of 'biological' weed control, as described in Chapter 6.

Bankside trees should be kept under control with a cycle of coppicing or pollarding before they create too much shade and leaf litter.

Pests

A true conservation pond is left to sort out a natural system of checks and balances between predator and prey. 'Ethically', species like heron, kingfisher, pike, perch, stoat and weasel should be left to their own devices to keep prey populations at an appropriate level, and water voles and water shrews should be allowed to undermine the banks if they wish. But certain alien species cause havoc in Britain and do need to be controlled before they annihilate native wildlife. The major problems of recent years have been mink and coypu.

The American mink, *Mustela vison*, has escaped or been released from fur farms. It became well established in the wild by the 1960s, especially in southwest England. Since then the species has spread to most of Britain and feral mink now live along our rivers, marshes and lakes. They swim very well (their feet are webbed) and they are entirely carnivorous, taking fish, small mammals, waterfowl, ground-nesting birds and domestic poultry. Like all mustelids, they are efficient killers but, unlike our native species, they tend to be wanton killers. They are only rarely taken by other predators (they are fierce fighters) except for humans. The usual control methods involve cage trapping and humane disposal by shooting, or approved strong spring-traps carefully placed in tunnels to avoid danger to other creatures. Alternatively, some means of fencing the mink out of reach of their prey must be found. The Ministry of Agriculture, Fisheries & Food and its counterpart in Scotland (DAFS) had an intensive trapping campaign against mink from 1965 to 1970, but then had to admit that the total eradication of mink was not possible.

With coypu, on the other hand, the eradication campaign does seem to have succeeded completely and this large herbivorous aquatic rodent is now thought to be extinct in Britain. It, too, was originally bred for its fur (nutria) and escaped to become a severe problem in East Anglia, causing substantial damage to the banks of rivers and ponds by its burrowing and also ravaging crops of root vegetables, brassicas and grain.

Amphibian Ponds in the Garden

With the disappearance of so many rural and village ponds, wetlands and heaths, garden ponds are becoming important sanctuaries for some forms of wildlife, especially the amphibians -- frogs, toads and newts — and some of the less common invertebrates. A garden conservation pond should aim to accommodate native species (especially endangered ones like the great crested newt and the natterjack toad) rather than decorative aliens, and aesthetics should make way for nature's demands, though the pond and its surroundings can still be very beautiful in an informal way.

Choose a sunny, quiet spot and give the pond as much space as possible — make it at least 4 × 2m (12 × 6ft). The great crested newt prefers an even larger pond. Smaller areas can be of value but will, of course, support much smaller populations. Frogs, for example, have been known to breed in old sinks sunk into the ground, but not in any number.

The pond need not be deep — perhaps about half a metre (18–24in) to make sure it does not freeze solid in winter to the detriment of any frogs hibernating in its depths. As with larger conservation ponds, there should be a variety of shape and depth: for example, frogs prefer to spawn in shallow water about 10cm (4in) deep on a broad shelf around the margins. The young will need an easy exit from water to land in due course, preferably on grass.

The lining of the pond can be pre-formed fibreglass, neutralised concrete (10–15cm or 4–6in thick) or plastic (preferably butyl) for a more natural shape, each needing a good layer of sand and soil as a planting medium.

Plant with native species taken from an unpolluted wild pond (after permission has been sought and granted). The plants and mud are bound to have invertebrates clinging to them, and perhaps even some newt eggs if the transplanting is in spring, so the pond will be off to a very good start.

For amphibian sanctuaries, oxygenating aquatic plants are important. Try one or two of the local varieties and see which of them thrive best in your pond. Newts wrap their eggs in broadleaved emergent plants like starwort and water soldier but also need some open water for mating. Tadpoles appreciate a little shade in hot weather, perhaps under waterlily pads, but all plant growth should be controlled to leave some open water in the pond. Blanketweed algae can be twiddled in ropes on a stick for removal if it becomes overwhelming, but check that any entangled tadpoles are returned to the water.

Plant a few marginals like kingcups, rushes (not reeds) and flag irises as cover for the breeding animals as they leave the pond, and as roosts for dragonfly nymphs when they are ready to become adults. However, leave some of the bank as rough grass which makes good cover for emerging young amphibians; but if the surroundings are to be lawn, cut it very short and very regularly from June onwards to deter young frogs and toads from loitering in the grass and getting themselves chewed up by the mower next time round. An area of rockery could be created around part of the pond: rocks are ideal for some amphibians, which will find perfect hibernation conditions and also sunning beds and shady retreats, and the rockery can be planted as an attractive waterside feature.

Like many ponds, even a brand new garden pond will soon be colonised by invertebrates and, once these are established, clumps of frogspawn can be introduced on the shallow marginal shelves — preferably from other garden ponds rather than raiding wild ones where frogs may not be common. If there is overcrowding among the tadpoles in due course (that is, if they are noticeably dying off), feed small quantities of rabbit-food pellets. The froglets will emerge from the water at about three months old just after a rain shower. Two years after the spawn is introduced the offspring will be old enough to return and breed in their home pond and the garden population will quickly build up.

Toads are slower to colonise new ponds and should not be hurried. They prefer a more mature, larger pond. Wait for two or three years after building a garden pool, then collect spawn strings and wind them around plants in deeper areas — say 15–30cm (6–12in). Toads are very faithful to the ponds in which they hatched and will walk long distances to return there for mating, braving many hazards on the way and frequently failing to survive them.

Newts are best introduced as adults in April. Use a pond net to catch perhaps a dozen (some male and some female) and then leave them to it. These will be palmate newts (on heaths and moors) or the common smooth newt. The great crested newt is a protected rare species and a licence must be obtained from the NCC before any are captured and transferred. The NCC or the British Herpetological Society will advise on sources; ideally two males and three females should start the colony in the first year, with another five or six being added the following year (like other amphibians, newts take two years to mature). If a pet shop offers great crested newts for sale, make sure that it has a licence to do so from the Department of the Environment.

Avoid putting fish into an amphibian sanctuary: they will eat all the tadpoles.

All these amphibians, incidentally, are the gardener's friends. They do an excellent job controlling snails, slugs and insects and indeed many gardeners keep pet toads in the greenhouse. They are friendly little creatures to have around.

Chapter 6

GREENERY

I'll hang my harp in a weeping willow tree,
And may the world go well with thee.
Anon, **There is a Tavern in the Town**

Algae

As plants are to the land, so algae are to water: they are the basis of the food chains and the main feature of aquatic life, be it in the oceans or the tiniest puddle or patch of mud. Some algae seem to be like bacteria, some behave more like animals, but most are classified as plants. Some are simply microscopic cells and others colonise to form visible blobs, plates and filaments, attaining maximum sizes in the great strands of seaweed. In fresh water they range from tiny cells to strings and ribbons more than a metre (3ft) long.

The major groupings are the green algae, the yellow-brown algae, the golden-yellow algae, the diatoms, the dinoflagellates, the cryptophyceae, the euglenoid flagellates, the red algae and the blue-green algae. The colour labels are often deceptive.

Within these groups there are many, many different genera, and species within the genera, and quite a few of them are common in ponds. The two main types of algal problems for pond managers are the 'blooms' of free-floating blue-green algae on the surface and the entangling strands of filamentous blanketweeds.

The more familiar pond species include the filamentous *Spirogyra* (like green hair floating in the water) and the notorious blanketweed itself (*Cladophora*) whose strands cause so many problems in ponds, blocking up outlets and entangling ducks and lines. Others form scums, like *Volvox*, a just-visible organism held together in jelly which congregates with others of its kind, or the minute single-celled *Chlamydomonas* species. The animal-like *Euglena* sometimes turns farmponds dark green; other species clog pond surfaces with layers of rusty red or yellowish scums, and a dark brown algae floats to the surface looking unpleasantly like clumps of sewage.

The planktonic blue-green algae, which are related to bacteria, are the ones that commonly cause 'blooms' on the surface, suddenly appearing in sunny weather so that most of the pond is carpeted in green, or bluish-green, or a brownish-purple perhaps, depending on the species. Diatoms, which tend to coat the pond bottom and its plants with a brownish slime, are special: they are very useful oxygenators of the lower levels, they are quite beautiful viewed through a microscope and, unusually, they produce a fatty oil rather than carbohydrates as a product of photosynthesis, and this is an important food for many small aquatic creatures.

Huge numbers of creatures depend on algae for food and for oxygen in the water. Most of the common pond algae are sunlovers and thrive in nutrient-rich water and, given both, their rate of increase is breathtaking — almost literally for other life in the pond at times, because if there is a sudden increase in nutrients in the water it will produce a sudden burst of algae growth which, if it rises to the surface, as many do, can block out light for other plants so that submerged aquatics cannot photosynthesise. Then oxygen levels at lower depths drop considerably as the shaded plants continue to respire, using up the available oxygen, whilst that produced by the surface bloom tends to be dissipated into the atmosphere rather than circulated through the water. Simultaneous eruptions of growth generally become simultaneous mass deaths as well, and when a large mass of algae dies its rapid decomposition by pond bacteria puts even more of a demand on the water's oxygen and in no time at all breathless fish are dying in large numbers.

Areas of pond can be kept clear of blooms and blankets by floating temporary rafts in sunny weather (without sunlight, algae activity is greatly reduced and the organisms tend to remain on the pond bottom) or by ensuring that no flush of nutrients from, say, fertiliser run-off enters the pond. (A temporary strawbale in the watercourse will help to filter out this kind of pollution, but the bale must be removed before it begins to decompose because in doing so it becomes a nutrient supplier itself.) The scum-forming *Volvox* likes a diet high in phosphates and low in nitrates, and it is often a lack of phosphates in the pond which is the regulating factor keeping a good balance of algae — enough to feed everybody and produce oxygen but not enough to cause harm.

It is important to remember that a good supply of algae in the water is generally of benefit to pondlife and that water with a very low algae population is as barren as a desert. If over-exuberant algae does become a problem, deal with it by mechanical methods rather than by spraying. Herbicides are far too sudden in their effect, producing an overload of dead matter which could seriously deplete oxygen levels in the water unless the debris is promptly raked out — an almost impossible task with most algae colonies as they break up into individual organisms which slip through all but the finest filters, though filamentous blanketweeds are easy enough to rake out or gather by twirling the strands on a stick. Herbicides are also indiscriminate: they tend to kill all plantlife (or all algae, if they are specific algaecides) not only in the pond but also downstream. If used at all, it should be under the supervision of the water authority and only in dull, cool weather when oxygen levels in the pond are high.

In the bad old days the instant response to algae blooms and blankets was to fill a bag with copper sulphate and tow it all over the pond behind a boat, or to throw in some potassium permanganate. Neither of these is now approved by the water authorities: copper sulphate can kill everything (including fish by asphyxiation) if doses are too high and purple 'pot perm' can be equally biocidal — it is after all used to purify water, which means to kill the life in it.

A useful management technique is to sprinkle limestone grit on the surface; it will drift down to the bottom taking the algae with it. Many blanketweeds sink of their own accord after rain or as the weather cools, after rising to the surface in warm, sunny periods, and they are usually no longer a problem by late August. The less light there is, the less active the algae, and it is worth establishing areas of plants with floating leaves (like waterlilies) as competitors for the available light.

All that growth, and the energy and protein that accumulates in living matter, really should not go to waste! Raked-out blanketweed can be composted and eventually used as a mulch but there are far more exciting possibilities for algae exploitation, and in some countries algae are specifically *cultivated*. Several studies have been made into the possibilities of supplementing livestock rations with algal protein, or of making organic manure for land crops.

In India, for example, the potential of blue-green algae is recognised as a biofertiliser and also as supplementary protein in food for livestock or indeed for people, if only it can be made acceptable to eat. (In Chad and Mexico, the cyanophyte algae known as *Spirulina* are drained through muslin and made into flat, dry cakes which are already quite an important food source for the local people.) Livestock feeding, which indirectly feeds the people, seems the most promising outlet for algae and many developing countries are experimenting with sun-dried algae food for poultry, fish and pigs. In Singapore, although algal protein is inferior in quality to that of soyabean, it has been found that up to 52 per cent of the soyabean meal in a pig-growing ration could be replaced with algae, or up to 30 per cent in poultry rations.

Recycling a resource, or its energy, makes a great deal of sense and a typical 'algae farm' encourages algae to grow in a pond enriched with animal waste (including human sewage). Substantial quantities of algae can be produced and at the same time the problems of sewage disposal are dealt with and full use is made of the nutrients contained in the sewage. The algae can feed livestock, or can be eaten by fish which are then converted into fishmeal for livestock, and the livestock excrete waste which is returned to the pond, which feeds the algae. . . and so on.

Harvesting cultured algae is the biggest impediment to its commercial exploitation and efforts are being concentrated on continuous filtration methods, or on centrifugation (which is expensive), or on 'flocculation', which means encouraging the algae to coagulate and float to the surface to form thick mats, so that the pond can be drained and the algae blanket swept up, heat sterilised and dried in the sun. The pondwater can be used for irrigation or for fish, and the sludge becomes manure which is spread on the land for crops which feed livestock and humans. Flocculation is encouraged by intense sunlight, high pH values and low nitrogen levels in the water.

The introduction of waste matter into the algae pond is often direct. For example, ducks on a pond automatically dung it, or chickens can be housed over the pond so that their droppings fall into the water through wire netting. Very often pigs are kept right next to the pond and their slurry is washed straight into the water.

In Thailand it is estimated that the sewage from a city of 100,000 people could yield 1,000 tonnes of algae in a year. It is practicable, given adequate sunshine, to produce more than 100 tonnes of algae per hectare (2.5 acres) per year with a protein content of nearly 50 tonnes — substantially more productive in terms of unit area than soyabean cultivation.

The whole point of these experiments is to treat wastewater products and at the same time to reclaim nutrients from the waste in the form of a usable product. Algae use the waste nutrients and sunshine to grow, and incidentally release oxygen, which is used in bacterial synthesis and biodegradation of organic materials, which releases nutrients such as ammonia and carbon dioxide for the algae to convert into protein. That is to say, the algae can convert the nutrients into forms which can be made available to higher plants (by manuring) and animals (by direct feeding). There can be *profit* in algae: never again should you look at blooms and blanketweeds as a nuisance! Half the dry matter in algae is protein, and they also contain vitamins. They have a high content of carotene, which does wonders for the colour of egg yolks and chicken flesh.

Several fish species can 'harvest' algae. For example, pig manure is used to encourage algae growth in a tilapia pond; the tilapia eat the algae, and the excess tilapia (they are notorious for population explosions) are fed to the pigs, which produce manure, which . . . and so on.

Algae also have a role to play in man's attempts to throw off the shackles of this planet and explore the oceans of space. French and Chinese scientists are using algae as all-purpose life-support systems for astronauts, providing food and, during photosynthesis, absorbing respired carbon dioxide in the spacecraft and replenishing its atmosphere with oxygen.

Aquatic Plants

For conservation ponds, or practical ponds which are also intended to encourage wildlife, the aim should be diversity: lots of different types of plants, and different species within the types, to offer the widest possible range of habitats for animal life. Plants provide food, oxygen, shelter from conditions, hiding places from predators, and platforms on which to rest, live, lay eggs and metamorphose. It is sensible to introduce local species which have already proved capable of thriving in the area, but it is also sensible to use plants which, while they are efficient oxygenators (a major role in water), are less rampant growers than nuisances like Canadian pondweed, an excellent oxygenator but it multiplies at an alarming rate and can choke up the pond. Instead, try plants like water milfoil, water starwort, water violet and water crowfoot.

Any planting up of a new pond must be done long before fish are introduced and, if at all possible, birds should be kept away to stop them uprooting the plants (protect them with chicken wire) while they are still vulnerable. Rhizomes can be transplanted, preferably during their dormancy, by chopping up clumps and selecting those on the edge for replanting elsewhere. Rooted plants need something to root into and if a new pond has a bare bottom some topsoil should be spread as silt, or the plants can be established in weighted, sunken pots of earth. Reeds are easily propagated by taking cuttings about half a metre long (say 2ft) and planting them half in water, with the bottom quarter of the cuttings in mud. Watercress is also very easily propagated from cuttings which can simply be dropped into the shallows, perhaps weighted by a small stone to stop them floating away, or pushed into the mud in up to half a metre (say 2ft) of water at the most.

Ideally the plantlife in a new pond should be allowed to develop naturally, brought in by birds, feeder streams or the wind. It will take longer than deliberate planting, of course, but only the plants that are suited to the pond's conditions will do well and that could save some unnecessary labour over mistaken choices of species.

Free-floating Plants

Free-floaters are either without roots or have fine, dangling rootlets which hang loose in the water and only very occasionally and temporarily latch onto the silt. They behave like algae: they can spread quite suddenly over the surface in a considerable mass. They are important oxygenators when reasonably controlled and are important, too, as food and shelter for all sorts of creatures. They can help keep the water looking cleaner because their shade deters algae suspended in the water or on the silt beneath them. Indeed the water under duckweed (the most common and abundant free-floating aquatic species) is often crystal clear and much favoured by fish.

Duckweed prefers still waters with a high content of organic matter: a well-populated duckpond or a cattle pond suits it nicely, and it is a favourite food of the ducks. Water soldier (which likes alkaline conditions) and the increasingly rare frogbit and water violet can also form dense carpets on the water and may need to be raked off now and then. Water hyacinth is a major problem in the tropics and subtropics, where desperate control remedies have been needed to prevent total blockage of waterways and ponds, and efforts are now being made to exploit this menace by harvesting it as food. It has been introduced to Britain as an ornamental aquatic plant, along with another tropical species, the water lettuce, but at present it has not found conditions so much to its liking that it becomes a menace. Unfortunately, however, many non-native species which are innocently introduced as curiosities or for their decorative appeal do become rampant and great care should always be taken.

In winter the new plants hibernate on the pond bottom, weighed down by stored starch, and float up again in spring as the starch reserves are used up and make them lighter.

Submerged Aquatics

This group of true aquatic plants are rooted in mud and live entirely under the water, supported by buoyancy and swaying with the current like grass in a breeze. Some of them may let their flowers rise above the surface. They are important oxygenators because, being submerged, they release all their oxygen into the water rather than into the air, and every pond needs to have adequate populations of submerged aquatics for that reason.

The very primitive stoneworts are almost algae but they keep their heads well down in the water and sometimes grow quite densely there. They are valuable pond plants and have been exploited as calcium-rich compost, as food and shelter for fish (they stink of fish!) and as water purifiers.

Canadian pondweed is a great oxygenator (watch it bubble) and is also a great reproducer, so much so that it became a real pest after it was introduced to Britain and Europe in the nineteenth century. It has fortunately now lost much of that excessive vigour in this country. Any broken-off fragment can form a new plant, and because it is so brittle there are always plenty of fragments.

Other submerged species which give shelter for tiny aquatic animals (and are therefore feeding stations for fish and many invertebrates) include the underwater mosses, water milfoil and water starwort, which sometimes floats free.

Floating-leaf Emergents

These semi-aquatic species all root in underwater mud, some deeper than others, but some or all of their leaves and flowers float on the surface of the water. Some have both submerged leaves and floating leaves, differently shaped, on the same plant. The floating leaves are useful for creating limited areas of shade and shelter for all kinds of pondlife and are good landing rafts for insects, many of which lay their eggs on the plants.

The group includes some of the pondweeds (*Potamogeton* spp.), and the shallows-loving arrowhead which has starchy 'nuts', sometimes fed to pigs. The amphibious bistort can (as its name suggests) live on land as well as in water, which means that it can survive drought and pond-draining. Then there are the species with showy flowers, especially the water crowfoot, with big white flowers, and of course the waterlilies which like sunlight, deeper water for their rhizomes, and no current.

Emergents

These plants protrude rather than float their leaves and flowers but still need to be rooted underwater in the mud. They live on the edge of the pond and include water plantain, which people in some parts of the world eat for its starchy base — and its other parts are enjoyed by animals and birds. Marestail likes the shallows, especially in calcareous ponds, but it can also live in much deeper water where it remains submerged. Water horsetail looks fairly similar, though it is not closely related, and can be as tall as 1.5m (5ft); it can grow in water up to a metre deep (3ft) and often forms extensive colonies.

Swamp Zone

Marginal plants in the swamp zone like to have their roots covered in water for most of the year and are tall enough to cope with rises in water levels. They can stand up to winds and waves. Their rhizomes tend to form solid mats so that a species soon dominates an area, especially the invasive bulrushes and the common reed, *Phragmites*. Then the mats gradually accumulate silt, beginning the relentless regression to dry land that is the destiny of every natural pond.

The main plants in this zone are reeds, rushes, sedges and horsetails and also plenty of flowering plants like flag irises, bogbean, the pretty little blue brooklime of chalky waters, water mint, bog arum and the bright, cheerful marsh marigold or kingcup, which used to be so common but now seems to be more of a childhood memory.

'Weed' Control

A pond plant only becomes a weed when it is in the wrong place or is too vigorous. The terms 'pondweed' and 'duckweed' are unfortunate: they suggest that the species are generally undesirable whereas in fact they are a vital part of the pond's ecology, giving food to creatures large and small. A living pond needs its weeds but cannot afford to be overrun by them.

As with algae control, herbicides should be avoided: mechanical weed control is preferable. A good biological control system might use grass carp, ducks and swans, all of which relish water weeds. Ducks eat just about any kind of aquatic plant matter down to about 45cm (18in) below the surface, so that the top remains clear but there is still plenty of food and cover for aquatic creatures. Their droppings fertilise the pond bottom to the benefit of most pondlife, directly or indirectly, though with too many ducks for the size of pond there may be fouling unless there is a free flow of water through it.

Swans, with their longer necks, graze on underwater weed to deeper levels and also keep reeds in check by devouring new shoots. They are highly recommended for keeping plants and algae under control, if they deign to visit and stay on the pond. Canada and other geese are useful bank grazers rather than pond weed controllers. Half a dozen Muscovy ducks on a hectare (2.5 acres) of pond will control duckweed. Of course, several species of duck and geese could also give you eggs and meat for the table as an added bonus.

Carp and some other fish control weeds by grubbing them out at the roots when they churn up the silt looking for food (which can make the w ter very murky and send lingering pond-bottom algae up to the surface): like pigs, they are born uprooters and 'roilers' but, unlike pigs, they are searching for animal food and do not eat what they have uprooted, so that the plant matter is left to decompose. Grass carp, however, are herbivores and can usefully control soft submerged weeds and algae. Goldfish will nibble away at filamentous algae, and silver carp are also algae eaters. Tilapia, if the water is warm enough, eat plenty of algae and weeds; crayfish also graze aquatic weeds and, like silver carp, grass carp and tilapia, they are good for the table as well.

If the activities of these biological controls need boosting, the main human efforts should be directed at raking off some of the duckweed (hard work but good results), scything emergents to control top growth (which will in fact ultimately *encourage* new growth) and grubbing out bottom roots at intervals if a plant is becoming too dominant. Cutting and grubbing are easier if the water level is dropped — but take care: a sudden change in the water level could adversely affect aquatic animal life, especially in warm or sunny weather.

Reeds and other silt-trapping marginals can be kept from invading the main pond if their area drops quite quickly into rather deeper water (more than a metre (3ft)). Submerged species will not on the whole grow at depths of more than 3m (10ft), and a well-designed pond includes deeper areas so that there will always be some open water (duckweed and algae permitting).

Reed control is an important part of pond management, and reeds can also be a valuable crop for thatching, weaving, fodder and even paper-making. Club-rush (which is actually a sedge) is a stubborn silter which must be controlled but it, too, has its uses for rushwork and its pith is used as a floating material in eel bobs and children's toy boats. Bulrushes are used in making cardboard, rope and woven items, and common rushes used to be valued for rushlights and animal bedding.

Free-floating duckweeds and algae can be removed with a boom dragged across the water to draw the mats to the side of the pond where they can be scooped out. An inverted garden rake or muck rake is another useful tool, or a good wooden weed rake with long, curved tines. Booms can be wooden battens with sacking attached, or lightweight plastic netting (small mesh size), dragged by two people on opposite banks or with the aid of a boat. On a small pond the weed can be hosed aside.

Cutting can be by scythe or with a cutter made of a series of steel blades with sharp, serrated edges. Each blade is twisted so that there is always a cutting edge at the stems, and they are linked together end to end with a chain at each end to weight the blades, and a rope attached to the chains for handling the cutter. It is hard work to use: it needs two people, usually one on the bank and one in a boat (with a third person controlling and steadying the boat), who use a sawing action across the water. If the area is accessible by wading, a draw hoe could be effective. For major operations, a mechanical digger or tractor can be fitted with a special weed-cutting bucket. For lighter weeds, a piano wire can be stretched between poles to cut plants near the base in early summer.

Work into or across any current or wind, otherwise the cut weed will simply pile up ahead of the cutter and make the job even more laborious.

Cutting has the advantage of instant clearance of the surface so that light can penetrate and the water is more open to atmospheric oxygen, but it will suddenly wipe out a large surface area of plant material which was home to a lot of creatures. It is only a very short-term solution, because it usually encourages rapid and synchronised regrowth, so that the problem is not only deferred but also increased later in the season.

For more thorough (and selective) eradication, rooted plants need to be dug or grappled out using a mattock or spade to chop through matted rhizomes, or a grubber (which is a strong fork bent at right angles). It is back-breaking work. Rampant reeds may have to be dredged out, which both removes the reeds and deters future invasion by creating deeper water.

Whether plants are grubbed, cut, raked or sprayed, the loose material must be removed from the pond rather than allowed to rot in the water. And then what should be done with it?

Some aquatic plants can be used, like algae, in wastewater treatment ponds to extract nitrogen and phosphates from pollutants and to treat sewage effluent so that dissolved nutrients can be recycled. Many water plants, however, are grown on purpose and harvested for food or for making paper and pulp, thatch, furniture, mats and baskets.

The Plant Harvest

There are several species of aquatic or damp-ground plants which can be harvested and some of them can form income-earning enterprises or at least supply useful material for home and farm.

Edible plants

Among the edible plants, some are only for those who like to experiment with food or who gain satisfaction from harvesting in the wild. Flowering rush, for example, has an edible root and some people also eat the roots of the white waterlily. The stems of the common reed (*Phragmites*) are said to be edible: the reeds are cut while green, dried and ground up into a flour which is so full of sugar that it toasts like marshmallow, and the American Indians use pieces of gum from the green stems as sweets. Several waterside plants are used as herbs, including meadowsweet, angelica, sweet gale (for flavouring beer) and water mint. Rice is a tropical plant which is not grown in Britain, but the most important waterplant crop in this country is watercress, which is cultivated intensively on a commercial scale in a few southern counties. It is an iron-rich plant which is used for winter salads at a season when fresh green leaves are rare. It is also a summer vegetable.

Growing Watercress

Watercress grows wild in streams all over Britain and can easily be grown for home use if there is a source of pure water, running preferably over chalk and about 0.3m (1ft) deep. It is readily established if the flow is slow, steady and above all clean. Simply push rooted pieces of plants into the shallows at the side of the stream at almost any time of year. The plant can be propagated by pulled shoots or cuttings of, say, 15–20cm (6–8in) of top growth (taken from a bunch bought in the greengrocer if you like) which can be scattered in the water in summer to root where they will if they are not washed away. Or the cress can be grown from seed in moist compost between April and August and planted out after a month or two.

Wild watercress still has a faint reputation for passing liverfluke to unsuspecting humans who eat its leaves, especially if it is growing wild by damp areas of pasture grazed by livestock. The fluke (a flat worm) is a parasite in sheep and cattle and passes part of its lifecycle latched onto a small species of water snail (*Limnaea truncatula*) which is common in shallow, standing water. At the next stage of its development the parasite is a cyst on plants.

Hygiene is therefore very strict on commercial watercress farms to ensure that there is no chance of the crop harbouring the fluke, and the harvested cress goes through some very thorough washing processes as well because the plant is eaten raw so that the parasite does not have to survive the processes of cooking.

Watercress is cultivated where the water springs from chalk or limestone, or in some cases Lower Greensand or perhaps light red sandstones. It needs a very high standard of water purity (good enough to drink) and quite a lot of attention right through the year if it is to succeed commercially. Today's growers use specially designed ponds to ensure an even water temperature (about 10.6°C or 51°F), a steady water flow and perfect water quality, which is constantly monitored. The usual system is a series of concrete-walled rectangular beds with gravel bottoms, each separately fed and drained (with independent inflow sluices) by parallel channels. They are on average about 9m (30ft) wide and anything from 35–90m long (120–300ft). The beds are emptied regularly and much care is devoted to the condition and grading of the bottom. It is a skilled enterprise, requiring plenty of attention to detail and considerable skill in packaging and marketing the product, packed in ice so that it arrives very fresh.

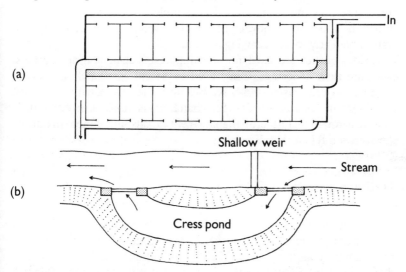

Figure 6.1 **Watercress bed**
(a) Double series.
(b) Off-stream pool controlled by sluices, for home-use watercress

Traditionally, watercress has satisfied a craving for fresh green leaves in early spring and it was said to cleanse the blood after the stodgy diets of winter. It is well supplied with vitamin C and has ten times as much vitamin A as the same weight of any other vegetable; it is also quite rich in iron, riboflavin, vitamin E and dietary fibre.

Harvesting Reeds and Rushes

Reeds have always been important for thatch in some parts of the country and are still a major fenland crop. In Romania they are farmed on a huge scale for a wide range of uses: pulped *Phragmites communis* (the common reed) is turned into cardboard, paper, cellophane and synthetic fibres, and any waste matter is converted into cemented reed blocks, compressed fibreboard, insulation materials, fertiliser, fuel and alcohol. About 125,000 tonnes of reed are harvested annually from the Danube reedbeds.

Reeds are also used in many countries for making windbreaks and fencing, or in basketwork or as firewood, and as mouthpieces in musical instruments. In some places reed marshes are utilised for treating sewage and wastewater. Reeds prefer high to medium levels of nutrition and some movement in the water, and need a raised watertable early in the growing season.

Various rushes are used for weaving mats, chair seating, baskets, bags and floor coverings. They used to be cut for thatching and for reinforcing wall-plaster. Both reeds and rushes are usually harvested biennially in rotation. Sedge for thatching is traditionally cut on a four-year rotation. Full crop management details for all these plants are given in the BTCV handbook.

Willows

Osiers are coppiced willows of various species and were used for basketwork (withies), fencing, eel traps and countless other applications in villages and on farms. They are exceptionally easy to propagate; they like a rich soil close to the water so that it is occasionally flooded, but they do not like standing in permanently waterlogged soil. Cricketbat willows are a long-term crop for a specialist market: there are only two or three merchants who buy it, and the trees need to be planted by clean, running water.

Aquatic Plants

As is the case with algae, people all over the world have made efforts to find a use for aquatic 'weeds', and in much the same way. They have been fermented to produce methane gas, or used as green manures (mulches, or ploughed into the soil), composts (wilted for a couple of days then piled with layers of soil, ash and animal manure to rot down) and animal feed. They are highly productive; their dry-matter content has 10–20 per cent protein in it and is often rich in carotene, iron, calcium and potassium, but their feed value depends on the conditions in which they are grown: best results are obtained in rich waters. Duckweeds are especially nutritious and rich in protein, and are much used in the tropics as food for ruminants and pigs (and sometimes humans, though they are not very palatable) as well as for manure. The main problem with aquatic weed crops is how to harvest them: they may be cut from the shore and hosed in, or cut by mobile harvesters, or even by people wearing flotation shoes! Then there is the problem of handling the harvested crop: the plants have a very high moisture content, which makes bulk handling and transport difficult. Yet their potential is too valuable to be wasted, especially when malnutrition is such an extensive problem.

Chapter 7

FISH AND OTHER POND LIFE

No human being, however great, or powerful, was ever so free as a fish.

John Ruskin, **The Two Paths**

Life began in water and it was a long time before aquatic organisms dared to venture onto dry land, let alone live on it as a preferred environment. Most ponds have an abundant population of aquatic animal life — microscopic plankton creatures, rotifers, worms, countless species of insects, spiders, snails, bivalves, crustaceans and, most apparent of all, fish and amphibians. Many of the animals spend their whole lives under water; others might be aquatic when they are immature but live out of the water as adults. A few have been exploited by man, mostly for food but also for their visual appeal or for their ability to regulate pests or keep the water clean. The relationships between such a wide range of animals in a natural pond are complex, but in a stable pond an ecological balance is generally achieved and maintained until some catastrophe strikes — more often than not wielded by the hand of man.

Fish

Fish need a reliable supply of dissolved oxygen, especially the mountain-stream species like trout which are adapted to cool, running water where oxygen levels are always high. All the game fish (trout, salmon and grayling) need clean, aerated water with a maximum tolerable temperature of 20°C (68°F). They also need clear, flowing, shallow water running over a clean gravel bed to be able to breed.

Coarse fish, which include all the other native species, are not nearly so demanding and can tolerate less oxygen, warmer water, and a degree of pollution which would quickly destroy their sporting cousins. However, each species of fish has its own preferences and Table 7.1 shows some of the ranges of temperature and water pH levels which they can tolerate.

Table 7.1: Fish details

Carp

In general
    Warm water, dense weeds, pH 5.5–9.0, can tolerate low oxygen levels
    Maximum tolerable temperature 38°C (100°F)
    Breeding temperature 23–25°C (73–77°F) (minimum 21°C (70°F))
    Optimum growth at 22–25°C (72–77°F), will grow at >13°C (55°F)
    Spawn on aquatic vegetation

Common Carp (*Cyprinus carpio*)
    Omnivorous bottom feeder
    Domesticated fish maximum size 120cm/30kg (4ft/66lb)

Crucian Carp (*Carassius carassius*)
    Omnivorous bottom feeder
    Maximum size 40cm/1kg
    Stunted variety can survive very low oxygen and short periods of freezing

Grass Carp (*Ctenopharyngodon idella*)
    Herbivorous — higher plants
    Maximum size 130cm/32kg (50in/70lb) in warm waters
    Cannot tolerate low temperatures

Goldfish (*Carassius auratus*)
    Omnivorous
    Maximum size 35cm/1.2kg (14in/2.5lb)

Golden Orfe (*Leuciscus idus*)
    Carnivorous (snails and larvae)
    Maximum size 60cm/3kg (24in/6.5lb)

Tilapia

Various species, mostly very warm water, a few slightly hardier.
Feed on plankton, plants, larvae.
Temperatures:
        lethal minimum 12°C (54°F)

optimum growth 20–32°C (68–90°F)
lethal maximum 40–42°C (104–108°F)

Trout

Require water of high purity, high oxygen levels, temperatures less than 20°C (68°F). Prefer clear flowing water, shallow gravel stream. Adults feed on flies, larvae, water insects, small animals.

Temperatures:

lethal maximum 25°C (77°F)
optimum growth 14–18°C (57–64°F) (12–15°C (54–59°F) brown trout) will grow 7–19°C (45–66°F)
spawning triggered at 6–13°C (43–55°F) (brown)

pH 6.5–8.0 (20–200mg calcium carbonate per litre of water
Spawn on gravel

Bream

Large lakes and slow rivers. Bottom feeder. Maximum 75cm/11kg (30in/24lb).

Perch

Predatory (hunting in shoals): fish, also larvae and crustaceans.
pH 4.0–9.2. Maximum 60cm/5kg (24in/11lb).

Pike

Lone predator: fish, amphibians, small mammals, birds, crustaceans, larvae, worms. pH 4.9–10.7. Maximum 150cm/35kg (5ft/77lb) (exceptional).

Roach

Algae, plant fragments, insects on plants. pH 6–8. Maximum 40cm/1kg (16in/2.2lb).

Rudd

Surface and midwater feeder: adult plants/insects, young feed on plankton. Maximum 30cm/1kg (12in/2.2lb).

Stickleback

Not fussy about environment. Will tolerate brackish water. Bottom feeder — larvae, etc. Maximum 10cm/30g (4in/1oz).

Tench

Densely weeded shallows; dislikes strong light. Bottom scavenger, useful for cleaning up pond debris. Will tolerate low oxygen. Maximum 60cm/6kg (24in/13lb).

All fish need food, and Table 7.1 also gives some idea of whether a species is carnivorous, herbivorous or omnivorous. The pH value of the water affects the abundance and variety of food in a pond and, generally speaking, acid waters contain less food than alkaline ones and therefore carry fewer fish which grow more slowly. Neutral or alkaline water with a pH of 7.5–8.5 is ideal for most fish.

While aquatic plants and the invertebrates that live among them are the main source of food for most fish, many also feed on small creatures that live on terrestrial vegetation and drop off overhanging plants, shrubs and trees. These water's-edge growers also give shade for the fish in warm weather. But the best sheltering places are provided by aquatic and marginal plants and by overhanging banks, deep lie-up pools in the bottom and plenty of nooks and crannies created by varying the contours of the bottom, the sides and the banks.

Individual fish tend to 'grow to the pond': the bigger the pond, the bigger each fish. A goldfish in a bowl never gets too big for the bowl, however much it is fed, but if it is released into a pond it will soon start growing. Therefore in fishfarming the size of the pond is as important as the stocking and feeding rates.

Conservation Ponds

Unless a pond is dedicated to the culture of one species, it is best to have as much variety as possible. Conservation ponds should be allowed to stock themselves naturally from feeder streams but if it is necessary to introduce fish artificially, then the WA's fisheries officer must be consulted first. No fish can be moved from one water to another without permission from the authority. Nor should fish be introduced unless the pond is capable of feeding them, which means that it must have well-established vegetation (including algae) and a good population of small aquatic creatures like plankton, insect larvae, snails and crustaceans. Most ponds soon gain stream tiddlers like minnow and sticklebacks, and probably some shoals of roach and rudd. These two species are superficially similar to the untrained eye but the rudd has a brighter shade of red in its tail and ventral fins.

One factor which governs the natural stocking of a pond is the competition for food, and this is where an aquatic environment has more to offer than a terrestrial one. Some fish are bottom feeders, foraging in or near the mud (for example carp, tench and bream), some are surface feeders concentrating on insects, others feed at various levels in the water. There is such a choice of food that at each pond level several fish species can co-exist quite happily, eating different foods at the same level.

Some of the bottom feeders, especially bream, tend to stir up the mud so much that the water is perpetually cloudy and this can inhibit plant growth. Other species of fish tend to overbreed (for example, roach) and if the pond is an enclosed one they will have to be culled every now and then before they become too dominant. Natural population controllers are pike and perch, both of which hunt other fish, but pike should be restricted to larger ponds where they tend to select diseased, dying or lethargic fish and thus keep the main population healthy. They are the big cats of the fish world, lying in ambush for their prey and then catching the victim after a quick sprint. Young pike feed on worms, crustaceans and insect larvae but as they grow older they turn to eating fish, amphibians, the occasional foolhardy duckling or water vole, and each other. The bigger the pond, the bigger the pike: exceptional individuals have reached 35kg (77lb) but anything more than 10kg (22lb) or a metre (3ft) in length is rare, and in ponds pike do not grow to much more than about 2kg (4.5lb).

Pike and perch are sometimes introduced specifically to control populations of farmed fish, but often pike in particular are seen as a nuisance and are trapped, sometimes by unpleasant methods. Perch are generally more tolerated, especially because they are popular for angling.

The species known as pike-perch, or zander, is not native to Britain though it has been introduced for angling. It is a member of the perch family with a fiercesome set of teeth and it, too, hunts fish. In favourable conditions it can grow to about a metre (3ft) long, weighing perhaps 12kg (26lb) — more the size of a pike than a perch.

Tench are all-purpose fish: they can be ornamental (there is a golden variety), they are stocked for anglers and they taste quite good if they are baked. They dislike strong light and prefer to scavenge in densely weedy shallows or to lie up in mud. They are also farmed: they breed quite easily in ponds (which some fish do not) and they mix well with carp if the ratio is about one tench to every ten carp.

Carp come in all shapes and sizes and colours, and are adaptable to conditions which would murder many other fish. They can survive perfectly well in highly eutrophic water with a low oxygen content, for example. Introduced to monasteries in Britain in the sixteenth century, they are valuable table fish, widely farmed, and a joy for the farmer with their rapid growth, but they need warmish waters: they breed when the temperature is 23–24°C (73–75°F). The domesticated common carp (*Cyprinus carpio*) is considered in more detail under Fishfarming (it includes the 'mirror' carp) and the gibel carp (*Carassius auratus*) is included under Ornamentals with its more familiar name, the goldfish. The crucian carp (*Carassius carassius*) used to be reared in artificial ponds but is now more often living wild: it is resistant to disease, and breeds very well if the water is warm enough for it. The crucian prefers clay soils; it is a bottom feeder and eats algae and plants, but also takes floating pellets. The grass carp (*Ctenopharyngodon idella*) is by no means native to this country but has recently been introduced to Europe and is useful for keeping aquatic vegetation under control. If food is in short supply in the pond, the grass carp enjoys a good ration of clover and lucerne. (The common carp will make do with boiled potatoes!)

Some of the carp family are highly ornamental, especially the fancy Japanese koi, which is a valuable fish and the target of thieves. It, too, is considered under Ornamentals.

Angling Ponds

Many conservation ponds (especially on farmland) are also angling waters for coarse fish. But if angling is the main concern, it is better to control the fish species and the population levels.

A new pond should not be stocked with fish at all until it has had a chance to settle and acquire a good density and variety of aquatic and marginal plants, and all sorts of invertebrates, as fish food. Give it at least a year: plant with fish in mind and encourae natural colonisation by animal life by introducing water and a little bottom material from local ponds. Small crustacea like shrimps and waterfleas can usually be bought from fishfarms if they are not found in nearby waters. Silt and weeds are assets in an angling pond as long as there is adequate open water for the anglers.

If the angling is not purely for private use, it will probably be necessary to obtain planning permission from the local authority, especially if there are approach roads, toilets, and anything that could be considered a nuisance to the neighbours. Coarse-fish angling can be profitable if the site is near a town and offers good facilities, including day tickets, whereas trout lakes are more profitable if the surroundings are particularly attractive and not so easily accessible. Game fish need special care and the pond must be designed with their requirements in mind.

The Southern Water Authority claims that leasing the fishing rights can give an income of as much as £15,000 per 0.4ha (acre) of water (though it is more typically £3,000–5,000). It may be possible to obtain a substantial grant from the Sports Council if a long-term lease is granted to an angling club — and the club will also make a contribution.

Have a good variety of coarse fish — for example, carp, bream, roach, rudd, tench, dace and perhaps perch and a pike or two — but avoid small fish like gudgeon, minnow and stickleback, though they make useful bait for anglers interested in predatory fish. Coarse fish spawn quite readily (usually about May or June), depositing their eggs on the bottom among stones and wood or on submerged plants. If conifer branches are placed in the pond temporarily during spawning, fish will be attracted to them as spawning surfaces and the branches can be removed, laden with eggs, if it becomes necessary to control the breeding population.

Trout Lakes

The native trout is the brown trout or 'brownie', but it is a mountain-stream species and may not do as well in ponds as the rainbow, which was introduced from the Rocky Mountains area towards the end of the nineteenth century. The brownie may possibly be induced to breed, given a good stream on a gravelly bed, but the rainbow is most unlikely to do so.

Rainbows are less expensive to buy in for stocking, easier to keep in still waters and grow to saleable size more readily. They are more self-assured than brownies, which means that they are less disturbed by hard fishing but are also more aggressive towards each other and to brownies, monopolising the available food and slowly ousting them from the waters.

Trout are territorial feeders and need plenty of variety in the pond landscape so that they can identify and possess an area of their own. They grow faster in hard water, especially on chalk and where the calcium content is greater than 150 parts per million because such conditions give optimum weed growth, and the weeds in turn shelter and feed plenty of aquatic creatures for the trout to eat. Young brownies and rainbows feed on insects (larvae and adults) and young crustaceans; adult brownies also live on insects and larvae and other small aquatic creatures, but larger individuals are like rainbows in that they eat other fish, including on occasion their own species. The pink flesh colour of some trout (which is less likely to taste muddy) comes from eating molluscs and crustaceans, whereas white flesh suggests the fish have been eating insects and plankton.

The main problems in managing trout lakes are ensuring that oxygen levels are always maintained (9mg oxygen per litre of water — which means keeping the water temperature below 20°C (68°F)), keeping pH levels stable (7.25 is ideal — in an emergency the balance can be adjusted by adding calcium carbonate at 500kg/ha (4cwt/acre)) and avoiding pollution at all costs. There must be ample aeration throughout the water, by means of cascades and circulation systems if necessary, with sluices to improve the flow.

Overcrowding may lead to disease and the stocking rates really depend on the potential for long-term food supplies, which means silt on the bottom and plenty of weed and invertebrates. With a new lake, limit numbers to about 250/ha (100/acre) and then add more in the second year if they grow well. It is cheaper to start with fingerlings (first-year fish) but it is easier to bring in two-year-olds perhaps 23–25cm (9–10in) long. Talk to the WA fisheries officer for advice on stocking numbers and sources of supply.

Trout lakes can be part of a general scheme on the farm. For example, a Devon dairy farmer converted some steep, boggy valleys into trout lakes because the land was not much good for the cows. He was also interested in conservation and landscape, and the lakes were carefully designed and planted to enhance the view as well as provide trout fishing. The passion for ponds grew and he created coarse-fishing ponds elsewhere on the farm, and then began to breed ornamental fish for sale. At present he still has a dairy herd, but who knows where his ponds will lead him?

Ornamental Fish

Some of the native species are very beautiful but their iridescent colours are subtle and are best appreciated at close quarters, or as a flickering shimmer as a shoal darts and turns in the water. Foreign species are often much bolder in their colouring, and in many cases this is because they have been selectively bred for colour. For example, the goldfish in the wild is a rather unimpressive olive, but the Chinese have bred it with great care for centuries, selecting those with the brightest colouring. However, the species can and does revert to its wild shade, especially if the breeding has been less than selective. The young do not develop their gold colouring until they are at least 50 days old, but colour is not the only factor sought by goldfish breeders. There are at least a dozen different varieties in Britain (goldfish were first introduced here in about 1705), some recognised by different colours and some by their shape. It is a resilient species and can tolerate heat and low oxygen levels. It grows fastest at a water temperature of 30°C (86°F). A sensible stocking rate for goldfish is a maximum of 15cm of fish per 0.1sq m (6in per sq ft) in due course.

The goldfish is a member of the carp family and several others in this group can be highly ornamental, especially the fancy strains of common carp developed in Japan since the nineteenth century — the glorious and very expensive koi carp. Carp are also important table fish and have been grown for centuries in China in simple earthponds heavily manured to encourage natural food for the fish. Domesticated carp, whether raised in China, Japan or Europe, become very tame and can live for many years, growing to considerable sizes. Individual fish respond to their owners (or certainly to those who feed them) and will come when summoned, apparently relishing a chat and a tickle; they are playful by nature. Some not only eat out of the hand but will even rest on the hand, as friendly as any other pet. Ornamental carp like the koi can be so valuable that their owners take out insurance cover and have quite sophisticated security measures to protect the fish from theft or harm.

The ide is a European fish but not native to certain countries, including Britain. There is an orange-red variety known as the golden orfe, often found in ornamental ponds, but to breed it requires deep, *clean* water. Other carp, however, breed freely, and those who have trouble increasing their stock of goldfish may find that the young are being attacked by water scorpions, great diving beetles and other invertebrates, or that the spawn are being devoured by the goldfish themselves.

There are a few points of general management which apply to ornamental or other fish kept for pleasure. They are often in rather small, shallow ponds in the garden and two major problems in these circumstances are overfeeding and ice.

In a frozen pond, there is nearly always room beneath the ice for the fish to hibernate. The problem is not that the fish freeze but that the ice traps toxic gases so that the fish are asphyxiated, though this does not often happen. Ice may also crack a concrete pond and this can be prevented by floating something slightly elastic (a rubber ball or a piece of wood) on the pond which can contrast and take the pressure off the concrete. A small area can be kept free of ice if an aquarium aerator is used to pump a little air constantly through the pond in cold weather, or the surface can be insulated with perhaps a sheet of polythene suspended on a frame a few inches above the water. Snow must be brushed off an iced pond to allow the light to reach plants. Take care that there is no run-off into the pond from paths or roads which have been salted.

Feeding should cease when the fish become inactive as the weather grows colder, and indeed they will be virtually dormant during the winter; start feeding again when the fish are seen darting about actively seeking food in the spring. Give them only as much as they will clear up in 20 minutes, otherwise wasted food acts as a fertiliser and the algae will explode into growth. In an emergency, trout and carp can be fed household scraps: trout like minced meat, and carp can have ground barley or flaked maize. Daphnia (water fleas) have been fed to ornamental oriental fish for centuries (old men would spend hours carefully hand-feeding their pet goldfish with reddish daphnia, one by one, perhaps thinking it might affect the colour of the fish's scales) and, if a quantity of daphnia are put in a tub of soft water, they will soon proliferate and make a useful larder of fresh food.

Eels

The life of an eel is long and venturesome. The European eel spawns in the Sargasso Sea, an Atlantic region near the Bahamas and Bermuda, and the larvae drift on the Gulf Stream for about three years, until they are carried towards the European edges of the ocean. Here they enter the river estuaries as elvers. The male elvers loiter around the mouth of the estuary but the females move up the river and live in fresh water for perhaps a dozen years, then they return to the Sargasso Sea to spawn and, it is thought, to die.

Eels are river fish but do quite often find their way into ponds, especially old millponds and the like where they used to be harvested in quite large numbers. They are almost mystical creatures, lying hidden by day in among tree roots, in clefts in the banks or bed, buried in the mud, or loitering near weirs and bridges or under the weeds. It is only at night that they become active and, on wet nights, they even come right out of the water and make their way across land to reach isolated ponds.

Freshwater eels live on larvae, snails, crustaceans and small fish. Their flesh has always been relished by humans and they have traditionally been caught in special traps, usually for eating at that stage. However, the elvers can also be trapped for rearing but they need plenty of protein and are not very efficient converters of food. They cannot breed in captivity so it is difficult to 'improve' the species as a domesticated fish.

Fishfarming and Aquaculture

With British agriculture under pressure, quite a few farmers are realising that water is three-dimensional and that by flooding their land they can in theory get a greater yield of saleable produce per acre — by farming different species at different levels in the water. A typical profile might have orfe at the surface, carp in the middle and crayfish (crustaceans) at the bottom. There is a growing interest in carp, and a minority interest in tilapia and tench, but the main fish species farmed in Britain are trout and salmon.

Rainbow trout for the table form by far the largest output of the fishfarming industry (brownies are only reared for re-stocking angling waters). Salmon spend only part of their lives in freshwater rivers and streams, migrating to the sea when they reach a weight of about 20g (less than an ounce), and farming therefore takes place on the coasts, or on the shores of brackish lochs, where the salmon are 'ranched' in floating cages.

Carp are not yet popular on the British plate though in many other countries they are a major table fish. In this country the common carp is farmed in southern counties (it grows fastest in water at 25–30°C (77–86°F)) and the fish are especially popular with Chinese and Indian communities in the UK, who prefer to buy them live. Grass carp are farmed on a very small scale, both as food and for controlling aquatic weed in drainage systems (e.g. fenland) and in fisheries – they can eat their own weight of duckweed daily.

Tilapia are a warm-water species, hugely prolific in the right temperatures, and the main tilapia scheme in Britain at present is probably Saughton Prison's production of fry for export, which is on a very considerable scale.

Ornamental fish are also 'farmed' for a substantial and increasing domestic market, especially for koi carp and goldfish. Other farms raise a variety of fish for selling to managers of angling waters.

Any form of fishfarming is necessarily beset by regulations and it is essential to seek advice before a new enterprise is planned. Talk to the WA fisheries officer, the Ministry of Agriculture, Fisheries & Food (or its regional equivalents) and the local council's planning officers about movement of fish, disposal of effluent, change of use of farmland and planning permission.

Trout farms depend on access to a reliable source of good-quality water, well oxygenated, with a dependable flow, free from pollution and running at a steady temperature. Thus the choice of site for such farms is limited by the water source. Trout need a water flow of one litre per minute per kilogram (0.2 gallons/minute/2.2lb) of fish at 15°C (59°F), for example: the flowing water maintains cool temperatures and water levels, oxygenates the ponds and flushes out waste matter. Fish screens and silt traps at the inlets are essential.

The smallest practical pond is about 100 sq m (1,000 sq ft). Anything larger than 500 sq m (5,000 sq ft) becomes a full-time enterprise and probably needs more than one person. Several small ponds are more efficient than one large pond: they can be used to separate fish of different sizes and stages; they can be individually isolated if there is a threat of disease or pollution; they are quicker to fill or empty; they are generally easier to maintain and to build. Most farms build a series of ponds but it is a mistake to have them stepped in a line with the same water flowing out of one pond into the next, taking its waste with it. Each pond must be able to be drained — for maintenance, for catching up the fish and for controlling disease.

Trout farmers rely heavily on artificial feeding, which is a major cost, and the fish are sold when they weigh about 280g (10oz), which they reach in about a year (longer in Scotland). Farmers of common carp, on the other hand, rely on natural algae and aquatic plants for their fish and raise them in well-manured or fertilised ponds (chicken manure is commonly used). The fish are harvested when the ponds are drained in autumn and winter, and they are sold at Billingsgate for twice the price of trout, or live to ethnic communities for double the dead-fish price.

Tilapia are said to be the fish that Jesus fed to the multitudes — and they do indeed multiply and grow at a phenomenal rate. The biggest problem in tilapia farming is that ponds are soon overstocked and some form of population control is necessary. They are essentially warm-water fish, native to Africa, Central and South America, and India, where they can tolerate poor water quality, partial drying out and disease: they are supreme survivors, and on top of their hardiness they can, in circumstances of prolonged daylight hours in hot sunshine, produce offspring up to eight times a year. The tilapia is sometimes dubbed the 'aquatic chicken'.

Outside their native regions, they can only be farmed if the water is heated. Winter temperatures must be above 10–12°C (50–54°F), below which they will die. Even the hardier types will not grow at temperatures less than 15°C (54°F). Perhaps the only possibility for rearing tilapia in Britain is to use ventilated polytunnels of the type used in horticulture (and increasingly for animal housing as well), enclosing an earthpond about 1–1.2m (3–4ft) deep and sited to receive all the available sunshine. They devour large quantities of algae and aquatic weed, and will also eat green vegetable leaves and tops, grasses and legumes (tied in bundles in the pond so that the stalks can be removed once they have been stripped), table scraps and ground grains. They could be usefully housed next to the kitchen garden!

Fishfarming can (and perhaps should) be combined with other enterprises like angling or the growing of fish stocks for other farmers (eggs, fry or fingerlings) or waterplants for fish ponds and ornamental pools. Irrigation reservoirs can be used for fishfarming, perhaps using suspended cages.

Or it could be part of one of the more elaborate cycles in aquaculture, combining the raising of pigs, ducks or poultry next to fishponds and using their manure to stimulate algae and aquatic plant growth on which the fish feed.

Crayfish

Britain's native crayfish were almost wiped out in the nineteenth century by a fungal plague, and it appeared again in 1981. Crayfish farmers therefore tend to import plague-resistant species, mainly from North America. As with any introduced species, precautions must be taken to prevent their escaping and breeding in the wild: there is no knowing what their effects on native wildlife and habitats might be.

The Institute of Aquaculture at Stirling University is a useful source of information.

Figure 7.1 *Crayfish*

Crayfish like clear, well-oxygenated and relatively alkaline water with perhaps 16–20mg of calcium per litre. The pH of the water should be more than 6, and preferably more than 7. They like a stony bottom with lots of hiding places — it helps to put in some brick rubble, bits of broken drainpipe, and so on. Water along the banks should be quite shallow, and clay banks will be used for burrowing.

They are omnivorous crustaceans and will eat weeds, algae, mosses, dead or dying invertebrates and fish, and any live animals they manage to catch. They also eat each other, which is why they need hiding places. They are at their most vulnerable just after moulting, a process which is essential to growth (they can only grow before the new hard skin forms). Juveniles moult perhaps half a dozen times in their first year and grow rapidly if the water temperature is more than 10°C (50°F). They mature at 3–4 years old (5–6 in northern counties) and then the males moult up to three times a year and the females once or twice. They can live for more than a dozen years.

Crayfish farming is a growing enterprise in Britain at the moment but there are no returns at all for the first three years. Anyone considering crayfish should seek advice from the British Crayfish Marketing Association, who can give details of management techniques and of co-operative marketing systems. Although there are few capital overheads in creating habitats for crayfish if local conditions are right, water quality is of vital importance and needs to be carefully assessed. The animals do not need supplementary feeding: they are scavengers and do a good job in keeping the water clean. They are harvested by selective trapping and they are sold live. At present the returns and potential look excellent.

Pond Crustaceans

The crayfish is the largest native freshwater crustacean in Britain. There are smaller ones, such as the freshwater shrimps, which are food for many fish and occur naturally except in very soft, acid waters. Like crayfish, they need calcium to build up a tough skin after each moult and, again like crayfish, they are most active at night and eat anything, especially decomposing leaves (elm, alder, sycamore), decaying dead animals and algae. Their preferences are similar to those of crayfish — clean, well-aerated water with hiding places in gravel or silt and somewhere to escape from strong sunlight.

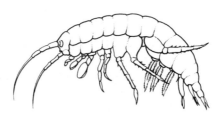

Figure 7.2 *Freshwater shrimp*

The freshwater hoglouse is very like its terrestrial cousin, the woodlouse, and is common throughout Britain. It will tolerate stagnant water and is happiest in organically rich ponds, delighting in sewerage systems.

The most abundant of all pond crustaceans is the waterflea, which plays an important role by feeding on microscopic bacteria and algae and being preyed upon by a wide range of carnivores from the tiniest animals to small fish.

Mussels and Snails

There are perhaps 70 to 80 species of freshwater molluscs in Britain, most of them natives, and they include water snails and bivalve mussels. As a group, they have useful roles to play in pond ecology:

they are sources of food for other animals (ducks and waders feed on them and many fish rely quite heavily on the tiny cockles); they tend to keep the water cleaner, and they are useful environmental indicators because they are sensitive to pollution. If water is quite badly polluted, there may be a large population of the *Pisidium* family but very few other molluscs; if there are no bivalves at all, the pond bottom is anaerobic, with no oxygen either because of unnatural pollution or because of a build-up of organic matter such as the substantial accumulation of dead leaves in a woodland pool. Dead mussels create pollution, too: their decaying remains can be fairly smelly.

Some species will only be found in ponds which have been reasonably stable for a long time, and their presence indicates a freshwater habitat which is good for diversity — an excellent conservation pond, in fact.

As the water of a pond becomes polluted or increasingly eutrophic, the balance of molluscan species within it begins to change and long-term observation and recording of the types and populations of each type in the pond will give a good indication of the pond's deterioration: there will be much less diversity of species. An ideal conservation pond for molluscs is free from pollution and its water is neutral or, preferably, alkaline. Lowland sites are generally much more promising than upland ones (they are on richer soils for the most part) but they are more susceptible to human pressures. Most of the molluscs do best where there is a rich, diverse population of algae and aquatic and marginal plants, and where water levels are fairly stable, though some specialists are adaptable to fluctuations. Some of the snails are wholly aquatic; others live on the pond's edges and marshes; and one or two mollusc species prefer brackish, fairly salty water.

Being shelled, they generally require a degree of hardness in the water because calcium carbonate is the basic material of a shell. The state of an aquatic snail's shell can give some indication of water quality: if the pH is high, the shell is usually shiny and smooth, but if the water is soft and acid the shell tends to be pitted, dull and brittle.

The bivalves, with their hinged double shells, are more sedentary than the water snails and are useful pond-cleaners, filtering the water to consume free-floating plankton. They are not edible, nor are pond mussels likely to harbour pearls, and this is their good fortune: it means the wild ones are not harvested to virtual extinction, which has been the fate of the river-dwelling pearl mussel.

The pond bivalves range hugely in size, from the numerous species of tiny 'pea shell' cockles to the much larger painter's mussel found in lime-rich water, which is perhaps 10cm (4in) across. But the most striking pond bivalve is the swan mussel, which can have a shell up to 20cm (8in) or more across. It lives on the pond bottom, often mostly buried in the silt, where it filters nutrients by drawing water in through one siphon, extracting what it needs, and pumping the rest out by another siphon — a miniature circulation system which oxygenates the animal's immediate surroundings. The hinged shells are shades of brownish-green and the colour and shape vary according to the environment — the depth of the water, whether there is a current, how much shade there is, and so on. Thus in the same pond there may seem to be several different species, whereas they may in fact all be swan mussels. They can live for quite a long time (perhaps a dozen years) and their age can be calculated from the ridged, concentric growth rings on the shells.

The mussel has an intriguing lifecycle and at one stage the tiny hatchlings (called glochidia) need to latch onto a fish, where they remain for a time as parasites. The animals seem to congregate in groups: in a pond survey you will find clumps of them here and there, especially where there is a little bit of current, which is useful to a creature that is virtually immobile and waits patiently for something to happen rather than chase off after food or mates. Chance seems to play a large role in the life of a swan mussel, and chance drifts on the current.

Snails

Snails, like bivalves, are molluscs. Nearly all the aquatic snails are herbivorous, rasping at algae and plants. The pond species include the trumpet snails, especially the ram's horn which is very good at controlling algae in an aquarium and is at home in large ponds. It has a typical child's drawing of a shell in a flattened coil and is commercially available for pond-stocking in a choice of colours — red, white or black.

The wandering snail is probably the most common pond species but the biggest snail in the water will be the great pond snail, with its spiralled shell ascending like a whipped icecream. It will eat decaying dead animals as well as algae and vegetable matter and can tolerate a degree of pollution. However, it could be a nuisance in a small ornamental garden pond where it tends to eat waterlily pads and other broadleaved plants for preference. To reduce the population, float a cabbage leaf in the water for a few hours or a day as a gentle trap for the snails to climb aboard.

Insects and Worms

Many ponds fairly throb with the activity of small aquatic creatures in various stages of their development. For most of them, adequate plant growth and algae are important — for some as food and for many as shelter and a surface on which to rest, mate, lay eggs, hide, hunt, pupate and generally live. Many insect larvae live in water: the caddis fly larva cocoons itself in an aggregate of sand, bits of leaves, twigs and shell; carnivorous mosquito larvae hang under the surface of the water to breathe. Dragonfly nymphs live in the water and are efficient hunters, capable of taking tadpoles and small fish as well as insects. Many fly larvae are aquatic — for example, those of midges, gnats, horseflies, alderflies and mayflies, the adults of which are often seen in great numbers near ponds, hovering in clouds, fluttering above the water, or clinging to waterside vegetation.

Leeches linger in the water, devouring worms and insect larvae or becoming temporarily parasitic on fish, snails or passing mammals, according to species (the horse leech does not in fact suck blood from horses: it swallows worms, snails, tadpoles and small fish). There are assorted worms living in the silt and mud of the pond bottom, cleaning up decaying matter in the shallows and for the most part found only where the water is not much more than a metre (3ft) deep. But down in the nether regions of the pond there may well be enormous numbers of tubifex worms, with their heads buried into the mud and their tails waving above them like a sea of red grass, fanning the water to gain a little more oxygen. They can survive in virtually anaerobic conditions and in a badly polluted, deoxygenated pond they may be almost the only form of life that survives, apart from a few persistent fungi and bacteria.

Adult insects need atmospheric oxygen, although they may live largely in the water. Most of the diving insects have breathing tubes which can be poked up through the water surface to replenish their oxygen supply, or have evolved methods of carrying air bubbles as life-support systems, and the water spider creates a diving-bell of air in which it can live. Most of the bugs and beetles in the water are carnivorous: great diving beetles, water scorpions and those upside-down backswimmers, for example, can tackle tadpoles or even small fish, though their prey is more likely to be something smaller, like moths; scavenger beetles, on the other hand, are herbivores and favour clumps of algae. Water boatmen look rather like backswimmers but are different in two major respects: they swim the right way up and they are largely herbivorous, favouring clumps of tiny algae and diatoms on the pond bottom.

Other insects live on the surface of the water. Water measurers seem to prefer stagnant ponds, where they take insects, mosquito larvae and waterfleas; pond skaters have a similar diet; whirligig beetles, too, make a meal of drowning insects.

A healthy pond should have a good variety of larvae, crustaceans, water beetles and bugs, snails, leeches and flatworms, and if pond-dipping shows that there are plenty of such creatures present, then the pond is likely to be unpolluted and suitable for fish and waterfowl.

Dragonflies and Damselflies

To many people, a pond is not a pond unless it is visited by hovering, darting dragonflies and their more dainty relatives, the damselflies. This group of creatures brings colour and restless movement to a pond, and gives a perfect excuse for sitting in the sunshine doing nothing more than pond-watching. They are essentially creatures of the sun and good memories.

There are perhaps 40 species of British dragonflies, some living by Highland bogs, others by slow rivers or fast streams, and many by ponds. The latter breed best where a warm, shallow, lowland pond lies on neutral or slightly *acid* ground (in contrast to the many creatures which prefer slightly alkaline conditions) and there will be a greater variety of dragonfly species near acid waters or on neutral clays.

The pond's wider surroundings are as important as the pond itself which should ideally be open to the sun to the south, with a fringe of emergent and marginal plants and a good rushy area. There should be plenty of sunshine and generally open surroundings, but also a sheltering belt of shrubs and woodland set well back from the water with rides and glades between the trees. They are territorial insects, especially the males of the larger dragonfly species, which protect their kingdoms by constantly beating the bounds while they hawk for prey. They can mate in flight over the water or will perch on marginal vegetation and the floating leaves of aquatic plants, and appreciate a pond with floating pondweed, waterlilies, bulrushes and bur-reeds.

Given a really warm and sunny summer, the adults might live as long as a month or two. Most species lay their eggs in slits cut into plant stems and leaves by or beneath the water, but the darter dragonflies hover on the water and deposit the eggs just under the surface. All the nymphs are aquatic, whether dragonfly or damselfly, and remain in the water for from one to three years, depending on the species, living on small animal prey (including in some cases small fish) and often eating each other as well. They will be preyed upon by fish and ducks, and will not do well if populations of these are substantial.

When the nymphs are ready for the final metamorphosis, they climb up a plant until their heads are just above the water, either at night or in the early hours of the morning, and a message seems to go out so that most of the nymphs emerge more or less simultaneously, climbing well clear of the water for the final transformation. A few hours later, the new adults are ready to fly but will not be in full colour for perhaps several days. Then they will flash and hover across the waters in a glory of rainbow bodies and dazzling, iridescent wings.

Amphibians and Reptiles

Amphibians and reptiles are cold-blooded creatures and they need the warmth of the sun to be able to function properly. Sunbasking is an important daily ritual and this need should be borne in mind if a pond is being managed with the interests of these animals at heart.

Amphibians

Amphibians spend part of their lifecycle in water and part on land, like some of the insects just described. The species native to Britain are newts (smooth newt, great crested newt and palmate newt), toads (common toad and natterjack) and the common frog.

Fish and amphibians do not mix well, to the disadvantage of the amphibians. Many fish eat frog and newt tadpoles (toad ones are not usually taken) and can eradicate the tadpole population in the confines of a pond.

Humans and amphibians do not mix very well either, again to the disadvantage of the amphibians, but not because they are eaten. The problem is human interference with the environment. Amphibians are generally loyal to their breeding ponds, returning annually to the pond in which they were hatched, very often to find that the pond has disappeared in the meantime or that the traffic, on a road which happens to cross the migratory route to the pond, is heedless of the marching armies of frogs and toads and they are slaughtered in huge numbers.

The garden pond is often the last resort for amphibians and the perfect pond is described in the Amphibian Ponds box (pp. 54–5).

Newts

The smooth newt is the most common one in Britain; indeed it is widespread in much of Europe, living in a wide range of habitats — mountains, heathland, marshes and woodlands. In spring, it needs standing water in which to breed; in winter it hibernates under stones or among tree roots. Mating takes place in the water and the eggs are laid individually on plants, hatching two or three weeks later. The tadpoles feed on insect larvae and small crustaceans, like the water flea, for three or four months; in most cases they become adults in August or perhaps a month or two later. They leave the water immediately and do not return to it until ready to breed at 2–4 years old. They feed on small crustaceans, snails, insects and larvae, worms and a few tadpoles, and can be lured towards the bank by a small worm dangled in the water on a piece of thread, wriggling enticingly as it is drawn donkey-and-carrot fashion towards the newt watcher.

Palmate newts are not found in Ireland or most of Scotland. They like sandy, chalky or peat areas, typically living on moorland and heath. Their breeding habits are similar to those of the smooth newt, and they feed largely on worms and insects.

The great crested newt is big, beautiful and rare in Britain and is now a protected species here. There is none at all in Ireland. Their habits are similar to other newts but beware of confusion: male smooth newts in their breeding dress also boast crests running the length of the back, and the females of both species have no crest! However, the male great crested newt has a more exaggerated crest and it is distinctly in two parts: a ragged body crest and a smoother tail crest with a definite 'waist' between the two, whereas that of the smooth newt runs in an unbroken wavy line the whole length of its back and tail, with some regularity to its serrations.

Toads

The common toad, well known to gardeners, can live for more than 40 years and can go without any food at all for several months if necessary. They hibernate in winter, buried in the soil under compost heaps and thick vegetation or hiding under stones and among tree roots, and they often hide up in similar places during the day, coming out at night or on damp, dreary days to feed on worms, slugs, insects, larvae, woodlice, snails and sometimes small frogs or young reptiles.

They return to ponds in the spring to mate in the water. The eggs are laid in very long strings wound around submerged plants and hatch after 8–10 days. The tadpoles develop into adults during the summer — usually in June and July — and the tiny toadlets, only twice the size of a thumbnail, take to the land, not returning to the breeding pond until they are four years old to mate in water perhaps 15–30cm (6–12in) deep.

The natterjack toad is found in western and central Europe, including England and southwest Ireland, but it is rare in Britain and is now a protected species. It can be identified by a yellow line down its back and by its gait, which is more of a scurry than a hop. It prefers sandy areas and often chooses brackish water for breeding in late spring and early summer. Its tadpoles and spawn are often devoured by the earlier-developing tadpoles of common toads and frogs, and the surviving toadlets are even smaller than those of the common toad when they emerge from the water.

An ideal natterjack pond is warm, shallow and sunny and its water has a pH value of 6 or more. The pond should be surrounded by at least a hectare (2.5 acres) of suitable foraging land.

Common Frog

The common frog is the most widespread of European frogs and is found all over Britain, but even the frogs are in retreat as their habitats dwindle. There are hardly any left in open countryside in some counties now, but they do seem to adapt very well to garden ponds.

Like other amphibians, the frog hibernates in winter (sometimes in the pond bottom) and mates in shallow water in the spring. The eggs are laid in clumps which float on the surface of the water in large jellified masses. Early in the season they may take as long as three weeks to hatch, and the tadpoles develop into froglets about three months after the eggs are laid. They often seem to emerge *an masse*, after a shower of rain, and the pond surrounds can be covered with a moving carpet of the tiny creatures. They gradually spread out into the countryside and will not return to the pond until they mature at 2–4 years old. A lot can happen to their pond in that time.

Reptiles

Britain no longer has native chelonians (tortoises, turtles and terrapins) but the yellow-splashed European pond tortoise did live and breed in this country in prehistoric times. It is now found in well-vegetated water in much of Europe and is often bought from pet-shops to live in British aquaria, along with the American pond tortoise and various terrapins like the Spanish, the Caspian and the Reeves. Some of these, especially the European, will survive here in warm, sheltered outdoor garden ponds, hibernating in mud near the water during the winter. They all like sunbathing on banks and islands but they hunt for their food in the water, during the evening and night, catching snails, worms, leeches, beetles, frogs, newts and occasionally very small fish, all of which they eat in the water rather than on land. Sometimes the European pond tortoise will lay eggs in Britain but they are usually infertile, though a few might hatch if they are kept in moist sand for a few weeks at about 21°C (70°F).

Lizards are terrestrial, but the common lizard likes damp places and will often take to the water to escape a predator. It is a very good swimmer, and may even dive down to hide in the pond bottom.

An even better swimmer is the grass snake, which is widespread throughout Europe but absent from Ireland and much of Scotland. The snake prefers to live near standing water, especially if there is a good patch of reeds and rushes. It slips quietly into the water and actively hunts its prey — mainly frogs, but also tadpoles, newts and a few fish. It is not venomous; it may strike out but with closed jaws, or may let off an unpleasant smell, and if it cannot escape its main defence is to 'play dead'. However loudly it hisses, it presents no threat to humans at all and can easily be tamed: it bears no grudge and soon responds to gentle, regular handling.

Grass snakes mature at three or four years of age and they may mate a couple of days or so after they come out of winter hibernation in April if the sun is warm enough. Usually they delay mating until May or June.

The eggs are laid later in the summer in a warm place like a compost heap or dung pile, and hatch between five and ten weeks later, depending on the temperature of the nest. The snakelets, perhaps 15cm (6in) long, feed themselves on worms and slugs, but are unable to hunt tadpoles and fish until they have learnt to swim, and generally hibernate soon after hatching.

Chapter 8

WATER MAMMALS
AND BIRDS

Four ducks on a pond, / A grass bank beyond,
A blue sky of spring, / White clouds on the wing:
What a little thing / To remember for years —
To remember with tears.
William Allingham, **A Memory**

Mammals

Most mammals can swim but only a few are well adapted for a partly aquatic life. They include the beaver (no longer native to Britain and disappearing fast elsewhere), the otter (also in retreat), the American mink (now feral in Britain, having escaped from fur farms, and capable of considerable damage to native wildlife), the water vole and the water shrew. The coypu, another fur-farm escapee which became a pest for a while, is now thought to be extinct in the wild, and there are no reports of invasions of other aliens like the raccoon, the muskrat or the capybara. The muskrat, a North American vole, has not been seen in Britain since before the First World War.

Water Vole

The plump, short-sighted water vole is endearingly oblivious to danger. It is a vegetarian and it sits happily on a mudbank chewing deliberately down an iris leaf as if it was a stick of Brighton rock. Often, however, the only trace of a water vole is a quiet plop as it slips into the water and glides away submerged. Its burrowing in banks can be extensive, with an underwater entrance and one or more exits above water as well. It is fond of strappy leaves like reeds and sedges, and also munches on willow leaves and those of some aquatic plants, and enjoys nuts, which it may store for winter. Its main food is grass, which it scythes on the bank and usually carries to the water's edge for eating.

Despite being preyed upon by many animals and birds, it is a successful mammal and is less disturbed by pond and river clearances than most creatures, moving back into its old habitat as soon as the vegetation grows again. Oddly, it does not like rain!

Water Shrew

The water shrew is probably just as short-sighted as the water vole but it is a better underwater swimmer and it will eat worms, insects, snails, frogs, small crustaceans and perhaps small fish. It has a shimmering appearance in the water, its fur spangled with silvery air-bubbles as it swims after whirligig beetles or dives for caddis fly larvae, and in winter it might be seen hunting under the ice.

Otters

Otters are river creatures rather than pond-dwellers and often range over many miles, using a river as the main highway of their territory and exploring its tributaries. They are now largely absent from most of England except the southwest, south Hampshire and parts of Norfolk, where they are present but not by any means abundant. They can still be found in Wales and Scotland, but are only in any numbers in northern Scotland.

Bats

Many bats choose to live near ponds, streams and rivers, where flying insects abound. The widespread pipistrelle, for example, can eat up to 3,500 insects every night in summer — which must be good news for pondside dwellers plagued by gnats.

The Daubenton's bat is sometimes known as the water bat and it favours alder-fringed pools where it skims low over the water, frequently dipping to catch an insect on the surface. It is sometimes accidentally caught itself by a startled evening fisherman.

The very silent, slow and steadily flying whiskered bat, often alone, likes to hunt over small ponds near woodland, while the distinctive long-eared bat often hovers hawklike over a sallow willow in full bloom in spring where the insects will be swarming.

Birds

Somehow, a pond without waterfowl is a dead pond, however much it may be teeming with life beneath the surface. Birds on the water are so *visible*: they move the water, they paddle and swim and dive and bathe, they fly in and out or clamber onto the banks, islands and rafts, and they add the dimension of sound — gossiping ducks, gabbling geese, croaking coots and piping moorhens against a sweet background of the songs of warblers and wagtails, the whistle of a kingfisher, and the occasional peeved expletive of a disturbed heron.

Then there is the busy time of mating and nest building, and the pleasure of seeing eggs hatching into brown and yellow ducklings and bright yellow goslings, scarlet-crested cootlings, soft brown moorhen chicks — colours of all kinds, and even more colour among the adults, whether the species is imported from exotic places or is native.

People can identify with birds. It is so much easier to recognise bird species and individuals, and to form relationships with them, than with fish (carp excepted) or swan mussels. Birds look you in the eye.

Different birds have different requirements of a pond but wildfowl are generally not too fussy as long as there is water, food, peace from predators or human disturbance, and somewhere to nest, to preen and to rest. Ponds become very social places: the ducks fly in from all directions in the evening for a good old chat and a feed, the Canada geese come planing in with loud honking and noisy greeting ceremonies, and at dawn and dusk the pond is alive with the sounds of calls, wing beats and splashdowns. During the day it is a little quieter: the ducks gossip among themselves and then have a vigorous, noisy bathing session about midday; the bold-faced coots have endless territorial furies against all comers, especially in the breeding season when even a swan can be chased off by a bullying coot, bringing to mind the Duke of Edinburgh's remark (in the context of carriage-driving) about swans being 'all serenity above the surface but paddling like hell underneath'.

Moorhens sneak in and out between overhanging branches, pottering around the inlets and edges of the pond, pecking with their heads and jerking their white-rumped tails as they swim or as they plod hesitantly over the grass on huge, carefully placed feet, and slipping silently out of sight under the water if caught unawares. A kingfisher darts suddenly from its vantage point, giving a brief whistle as it flashes low over the water seeking minnows and sticklebacks, and always lifting the human spirit with its tropical colouring and promise of halcyon days. The heron stands motionless, knee-deep in the water, waiting patiently for a passing fish which, more often than not, the big bird misses with its stab and eventually, lazily, it rises into the air with a shrug, as if it did not care, and flaps away on leisurely wings.

The different species have different feeding niches, hence the variety on a good pond. Mallards and shovelers dabble the shallows and also eat plenty of duckweed. Pochard and tufted duck are divers: they feed deeper than the dabblers. Tufted ducks are a favourite target for the spluttering coot, which seems to react to their black-and-white colouring as something akin to its own, and it paddles and flaps with great vigour and speed across the water to drive them off its territory. The ducks dive at the last minute and the coot, clucking angrily to itself, loses its target.

Coots dive quite deeply for algae and waterweed, which they feed to their nagging young for several weeks. Moorhens are also long-suffering parents, and in fact the offspring from an earlier hatching help take care of the new chicks, often working harder to feed them and chivvy them than the parents. Moorhens prefer several small pools: they become very territorial in the breeding season and those hard-working teenagers are soon squawking in indignation as the parents chase them away to find their own patch elsewhere, perhaps in a quiet, shallow, woodland pond which was once a millpond silt-trap.

Wigeon and geese graze the pond's surroundings; mallards, too, wander over the grass for a bit of grazing and love nothing better than a temporary rain pool on a lawn where they dibble with vigour. For most ducks a pond averaging perhaps 0.5–1.2m (say 2–4ft) in depth is adequate, with some parts down to 2m at most (say 6–7 feet) and plenty of weedy shallows and marginal cover. There must be shallow areas of bank to enable ducks to leave the water with at least some dignity: they are far too waddly to jump.

Wildfowl help to manure a pond and at the same time help to control its plantlife, but if the wildfowl population is too large there are several ways the pond can suffer. The banks may be trodden to a grass-bare mud; the aquatic plants might be overgrazed or uprooted if they are not well established; excessive manuring can quickly lead to deoxygenation; and in extremely crowded situations botulism can be a problem, especially in hot weather. The organisms live in the mud in stagnant water, where they feast on decaying vegetation, and caused major headaches to the birdmen of St James's Park in London.

The royal parks have very careful breeding programmes for their ornamental waterfowl in the London area, making sure that inbreeding is minimised, yet they seem to fail with those delightful tree ducks, the mandarins, which usually breed easily enough and raise their young successfully in the wild. Mandarins are natives of eastern Asia but have established themselves in the wild in Britain. The drakes are even more showy than the closely related Carolinas from North America but the ducks, like all female ducks, are quietly dressed for good camouflage on the nest, and they nest at heights — literally in trees in some cases and often well away from water. The young 'parachute' down to the ground, usually without harm. Many duck species nest a long way from the pond and the young have a trek to the water the day they leave the nest.

Tree-nesting ducks appreciate man-made quarters — perhaps a wooden house on a stilt in the pond, with a ramp down to the water for the ducklings, and they have been known to take over boxes put up for barn owls. A barrel on its side in the fork of a tree serves mandarins well as a nesting box.

Most ducks lay before ten in the morning and gradually build up a full clutch over several days. The first three or four eggs lie exposed, but thereafter the duck always covers the clutch with plant material whenever she leaves the nest. It will be difficult to detect even when you know exactly where she has laid. Wild ducks are careful not to betray the site; she will choose a very indirect route back to the nest after coming off to stretch her legs for a feed and a chat, however much she trusts the person from whose hand she has just eaten grain. When she is sitting, only her bright, watchful eye gives her away, and if the nest is approached too closely she will hiss like a snake — which is an effective deterrent when it is unexpected.

Ingenuity is boundless when it comes to devising duck houses and nesting boxes for wildfowl. The aim is to give the bird and her eggs security from predators, of course, and a feeling of being hidden. A natural-looking artificial nesting site in a conservation area could be made by driving short poles into the pond, in a circle, and bending them at the top to be tied wigwam fashion; the frame can be covered in wire netting thatched with rushes, reeds, long grass, and so on, with a floor of similar materials and a small entrance on the north side. If it is on an island protected from rats, foxes and cats, so much the better. Readymade duck baskets are a more elaborate version, woven into a bottle shape so that they are enclosed except for the entrance hole, and they can be set on stakes over the water or on rafts. Give ducks time to get used to them, and thereafter they will be occupied every year. Ducks are very faithful to a good site.

Provide a few low, roofed shelters as refuges from hot summer sun or cold winds and frosts. They can be very simple, no more than a gently sloped roof on a low frame, and will be especially appreciated by ornamental ducks which are confined to their own pond. Confinement is necessary more to keep predators out than the ducks in (non-indigenous species must be wing-clipped or pinioned). A pond for ornamentals can be small — perhaps 4m (13ft) across, shelving to about 0.75m (2ft 6in) deep at the centre. But an area of water needs to be kept open somehow if the pond ices over.

Mallards often start laying ridiculously early, even in February, and continue ridiculously late, even up to Christmas, but such out-of-season clutches are unlikely to hatch well, or to survive if they do hatch. The ducklings are self-sufficient; they are not fed by the adults and they need a good supply of insect life to give their fast-growing bodies adequate protein, so that in colder months they are doomed from the start.

If food is offered to wild mallard, the message soon spreads and the population builds up quickly. In my own experience, one nesting mallard on a millpond can become 20 or 30 recognisable and tame individuals within the year if they are fed regularly, and up to 100 or more when other ponds are frozen. The advantage of an old millpond in winter is that at least some of the water remains open where the current flows; and the ducks will insist on lining the edge of the ice in Indian file, occasionally slipping accidentally into the water as the ice rim gives way beneath them.

Domestic ducks kept for eggs or meat can do well in ponds as small as 2m (6ft) in diameter and only deep enough for 'ducking' (the midday splash-around, which is similar to a blackbird's bathing but communal and much noisier). Geese like a pond but spend much of their time on the land and the domestic types will graze happily enough alongside sheep. Unfortunately, goose droppings seem to be more effusive than those of sheep and are certainly more messy.

Of course the most majestic of 'ornamental' waterfowl are the swans, if they deign to make use of a pond. They need space — not just water space, but air space, on the principle of airports. A pond surrounded by trees gives a swan no room for landing and taking off, and such a pond will always be rejected by prospecting swans. A clear runway is essential, on the water and over the surroundings. I once watched a young swan, who had arrived on one of my ponds because it was too exhausted to fly further, trying to get airborne for almost a month before he partially succeeded (he only made it as far as a neighbouring walled garden, from which he had to be rescued). There were tall alders on most sides of the fairly large pond, and where the trees parted the ground rose quite steeply up a hanger. The swan tried to gain runway space by making his initial takeoff in anticlockwise circles on the outermost edges of the water, flapping to get airborne as he paddled hard, day after day. It was sad when he eventually succeeded and he never returned, although he had been well fed and cared for while in residence, despite being bullied by the coots and maternal mallards. A wild swan, however tame it becomes, needs to know it can fly free when it wants to do so.

And then there are cormorants. They sometimes fly quite a long way inland and are not warmly welcomed at fishfarms and angling ponds. However, they often come in without any real intention of feeding; they will rest awhile and then move on. In some parts of the world cormorants have been trained for at least 2,000 years to catch fish for their owners: they are tethered to a boat on a long line and are prevented from swallowing the catch for themselves by means of a grass ring around the throat which only allows them to gulp down the smaller fish. They are intelligent birds; after a month or two of careful training a wild cormorant learns to respond to the fisherman's whistle and will return to deliver the catch even when there is no restraining cord to draw it back to the boat. Many will go on working for ten years or so.

A much smaller fisher, and far more selective, is the busy, plump little brown dabchick which works methodically near the pond margins, suddenly diving for stickleback and disappearing for so long on its underwater hunt that its presence is forgotten before it pops up again a long way beyond.

Most ponds, being so attractive to insects, are in consequence attractive to a wide variety of birds. Swallows, swifts and martins swoop and skim over the water, harvesting the insect swarms. Warblers live secretly in the reedswamps and the willow carr, betraying themselves by song. Reed buntings fly bouncily about their business, or perch for a while with their tails flicking. Colourful grey wagtails bob on the shores and stones within hearing distance of a weir or a cascade, only really content if they are near tumbling waters that move as restlessly and ceaselessly as they do themselves. With their rocking tails and quick little scurries on nimble feet, the bright greys and yellows of their plumage are almost a match for the true yellow wagtails that only come as summer visitors and then, like so many of the songbirds, head for the warmth of Africa when Britain faces its winter.

Chapter 9

PONDS FOR PLEASURE

*The wise man delights in water; the good man delights
in mountains. For the wise move but the good stay still.
The wise are happy but the good secure.*

Confucius

Pond-watching is the bonus which compensates for all the labour of creating or rescuing or caring for a pond. Henry David Thoreau, the nineteenth-century American philosopher, naturalist and writer who inspired Gandhi, passed three years doing little more than pond-watching by the wide waters of Walden and found that, maintaining himself solely by the labour of his hands and simplifying his needs to the most basic, 'by working about six weeks in a year, I could meet all the expenses of living'. For the rest of the time he watched and wondered, and grew wise under the influence of the pond, living on its edges in a log shack he had built for himself with the help of a borrowed axe. The chosen interlude gave him an opportunity to know himself and also to put human needs and aspirations into perspective. He died ten years after leaving Walden, at the age of 45.

Walden was a natural pond but it was not out in the wilderness: it was less than an hour's stroll from Concord, Massachusetts. Yet it gave the young man great peace and delight. Some ponds, of course, are designed purely for pleasure, perhaps to catch the eye with a gleam of water set as the focal point of a sweeping vista viewed from a country mansion, or to give city dwellers an oasis at the heart of the everyday hubbub. Many ponds are for private enjoyment in a small garden, lending it an extra dimension of light and movement. Water is life and light: it beckons.

In Japanese gardens, water is used more precisely and with exquisite artistic skill; its cascades are formal, its plants all perfectly formed and controlled, and one feels like a clumsy intruder in such a delicate and graceful setting.

166

In China, however, ubiquitous water is part of more robust gardens which have a sense of being lived in and enjoyed by humans who laugh, tell jokes, relax, have family parties, discuss philosophy and art, argue and have a strong sense of profundity, creation and wisdom.

Our own Western gardens today owe more to the Chinese than is perhaps realised. For centuries the great European gardens were decidedly formal: they were laid out by geometricians, with straight lines and clean circles that attempted to control the exuberant wildness of nature. Precise, organised patterns set the tone, and flowers and trees were forced into compliance and made tidy. The Chinese, on the other hand, adapted their gardens to nature's flowing, informal contours; they did not seek to tame nature but to be in harmony with it, yet in reality they were exceptionally artful.

Such gardens were described in letters written by the Jesuit, Père Attiret, in 1749, and by the end of that century the English, in particular, had taken the Chinese 'spontaneous irregularity' to heart. Indeed Alexander Pope had already started such a fashion 30 years before the Jesuit's descriptions; he had been inspired by Sir William Temple's essay on oriental gardens and had transported China to Twickenham, where he created a garden in which 'Spontaneous Beauties all around advance' and where he never lifted a finger without first consulting the 'Genius of the Place', which meant that he would always look first at the natural water levels and contours and then design the gardens to enhance and fit in with what was already there. The geometrician had no place in such a setting.

No Chinese garden was truly haphazard; each was most carefully designed to appear natural. The quiet streams that meandered around gently curved foothills, the paths that wandered so that their destination was always around another bend, the rocks strewn here and there as if the giants had scattered them at random like pebbles falling from a schoolboy's pocket, the flowering fruit trees and slender willows that grew where they would, or leaned longingly over the water: they all seemed so natural but were as planned and contrived as the numerous bridges and island pavilions.

Water and rocks were essential to the Chinese garden, whether in the complex of lakes, pools and linking rivulets of a huge imperial estate or a tiny backyard rockery with a trickling cascade. Water and rock are complementary: the rock is hard, bright and strong and the water yields and reflects; but the rock is passive and static, while the water is active and dynamic.

Many Chinese gardeners made the most of water's movement: they loved the sound of springwater falling into a pool, or cascading between rocks, or gurgling as it flowed among small stones and pebbles, though they never liked fountains, which forced water to act unnaturally.

But on the whole it was the peacefulness of water that appealed most strongly to the Chinese. They associated it with the misty mountain lakes that merged imperceptibly with the clouds, giving an impression of unlimited space, serenity and wholeness. The gardeners knew, however, that in gardens the water needed to be set off with little features which would enhance the water's beauty, and would in turn be enhanced by it, like a naturally lovely woman made lovelier still by a flower in her hair.

Thus the Chinese water gardeners made islands and promontories, and linked them with bridges and causeways, and set elegantly simple buildings and carefully chosen trees to be reflected in the water. Everywhere there were curving lines and random irregularities rather than the formal lines of Europe (the Chinese believed that evil spirits could only travel in straight lines). Ponds had hidden backwaters, little inlets disappearing from view behind a rockery bank, a pavilion peeping round a tree, a wall's low arch spanning the water, barring human progress but letting the water pass and deceiving the wanderer into believing that the water wound on forever beyond the wall, though in reality the pond's boundaries were just out of sight on the other side. A walk through the gardens was always an adventure.

Reflected images are the essence of many Chinese gardens: bridges, rocks, grottoes, trees, magnolias and waterlilies, walls, pavilions, changing skies, the moon and the stars. Yet virtue can be seen even in cloudy waters, which are likened to pearl or jade. Oh precious algae!

Students, scholars, philosophers and lovers would loiter in the oriental gardens, tempted to explore the zigzagging bridges and the pergola causeways draped with wisteria, which let them walk among their waters and feel as much a part of the scenery as if they were depicted on a willow-pattern plate, while we remain on the shores of our ponds, only watching. They built pavilions that invited the visitor to linger and gaze deeply into the lake for hours at a time, conversing silently with the carp or catching the moon's reflection in a cupped handful of silvery water. (Wiltshire's 'moonrakers' were no such romantics: their catching of the pond moon was only an excuse to cover their smuggling.)

Water gardens, then, are places for reflection and contemplation, where idle hours drift by while the soul is soothed and memories of aching muscles fade away. The setting need not be formal, nor a garden, nor even quiet, however. Ponds can give more than solace and a living pond, attended by conversational ducks and teeming with underwater life, can distract the willing for hours. There is always something to notice, something to watch, something to look for in the water or the mud, and there are always small insights to be gained, not just into the nature of the pond but into the nature of life. They say that those who live near water are calmer, more at peace with themselves, soothed by magical ions and inspired by their musings. Perhaps that is sometimes true but they are also invigorated and refreshed by the exuberance of life that is the reward for those who have understood, cherished and laboured for a pond.

PLANT TABLES

Amphibious Bistort	*Polygonum amphibium*	EM M–E
Alder, Common	*Alnus glutinosa*	T
Arrowhead	*Sagittaria sagittifolia*	EM E
Blanketweed	*Cladophora* spp.	FF M–E PT
Bogbean	*Menyanthes trifoliata*	MARG O Acid
Bogmoss	*Sphagnum* spp.	MARG O Acid
Brooklime	*Veronica beccabunga*	M–E
Bulrush (Reedmace)	*Typha latifolia*	EM M–E
Bur-Reed, Branched	*Sparganium erectum*	EM M–E Clay
Bur-Reed, Floating	*S. angustifolium*	FL O Acid
Buttercup, Creeping	*Ranunculus repens*	MARG
Clubrush, Common	*Scirpus lacustris*	EM M–E
Clubrush, Floating	*S. fluitans*	O Acid
Cotton Grass	*Eriophorum* (sedge)	MARG O Acid
Duckweeds	*Lemna* spp.	FF M–E
Frogbit	*Hydrocharis morsus-ranae*	FF E
Great Bladderwort	*Utricularia vulgaris*	FF G–M Insec
Great Water Dock	*Rumex hydrolapathum*	MARG E
Hornwort	*Ceratophyllum* spp.	FF M–E Alk
Horsetail, Marsh	*Equisetum palustra*	EM O–M
Horsetail, Water	*E. fluviatile*	FL O–M
Iris (Yellow Flag)	*Iris pseudacorus*	MARG O–E
Marestail	*Hippuris vulgaris*	SUB M–E
Marsh Marigold (Kingcup)	*Caltha palustris*	MARG
Marsh Yellowcress	*Nasturtium palustre*	MARG
Meadowsweet	*Filipendula ulmaria*	MARG
Pondweed, Bog	*Potamogeton polygonifolius*	O Acid
Pondweed, Broadleaved	*P. natans*	FL M-E
Pondweed, Canadian	*Elodea canadenis*	SUB M–E
Pondweed, Curly	*Potamogeton crispus*	SUB M–E Clay
Pondweed, Fennel	*P. pectinatus*	SUB E PT
Pondweed, Red	*P. alpinus*	SUB M Acid
Pondweed, Shining	*P. lucens*	SUB E Alk
Pondweed, Various-leaved	*P. gramineus*	SUB O–E Acid
Quillwort	*Isoetis* spp.	SUB O–M Acid/Pl
Reed, Common	*Phragmites communis* or *P. australis*	EM M–E

Reed (Sweet) Grass	*Glyceria maxima*	EM E Alk
Rice Grass	*Zizania latifolia*	EM
Rush, Bulbous	*Juncus bulbosus*	MARG O Acid
Rush, Flowering	*Butomus umbellatus*	EM M–E Clay
Rush, Jointed	*Juncus articulatis*	MARG O–E Acid
Sedges	*Carex* spp.	MARG
Shoreweed	*Littorella uniflora*	SUB O–M Acid
Spearwort, Lesser	*Ranunculus flammula*	EM O–M Acid
Starwort	*Callitriche* spp.	SUB O–E
Stonewort	Algae, e.g. *Nitella* and *Charavulgaris* spp.	SUB O–E
Sweet Flag	*Acorus calamus*	EM M–E Alk
Water Celery	*Berula erecta*	MARG M–E Alk
Water Chestnut	*Trapa natans*	Alk
Watercress	*Nasturtium officinale*	EM M–E Alk
Water Crowfoot, Common	*Ranunculus aquatilus*	SUB O–E
Water Crowfoot, Pond	*R. peltatus*	SUB M–E
Water Fern/Moss	*Azolla filiculoides*	FF E
Water Hawthorn	*Aponogeton distachyus*	FF
Water Hyacinth	*Eichhornia crasspies*	Tropical
Water Lettuce	*Pistia stratiotes*	Tropical
Waterlily, Fringe	*Nymphoides peltatum*	FL E
Waterlily, White	*Nymphaea alba*	FL M–E Pl
Waterlily, Yellow	*Nuphar lutea*	FL E
Water Lobelia	*Lobelia dortmanna*	SUB O Acid
Water Milfoil	*Myriophyllum* spp.	SUB O–E
Water Milfoil, Alternate-leaved	*M. alterniflorum*	SUB O–E Pl
Watermint	*Mentha aquatica*	MARG M–E Alk
Water Plantain	*Alisma plantago aquatica*	EM M–E
Water Soldier	*Stratiotes aloides*	FF M–E Alk/Pl
Water Violet	*Hottonia palustris*	SUB E
Willow, Crack	*Salix fragilis*	T
Willow, Cricketbat	*S. alba coerulea*	T
Willow, Osier	*S. viminalis*	T
Willow, Sallow	*S. caprea*	T

Key to plant tables

FF	Free-floating
FL	Floating-leaved
SUB	Submerged (but may have floating or emergent flowers or some floating leaves)
EM	Emergent
MARG	Marginal
T	Tree

Acid	Prefer or tolerate acid waters
Alk	Prefer alkaline waters
Clay	Usually found in clay areas
Insec	Insectivorous
E	Mainly eutrophic
M	Meso-eutrophic or mesotrophic
O	Oligotrophic
PT	Pollution tolerant
PI	Pollution intolerant

Some Plants for Ducks and Other Wildfowl

	Food	Cover	Duckling cover
Amphibious Bistort	x		x
Arrowhead	x		x
Bulrush			x
Bur-Reed, Branched	x	x	
Buttercup, Creeping	x		
Clubrush	x	x	
Duckweeds	x		
Great Bladderwort	x		
Great Water Dock	x		
Hornwort	x		
Horsetail, Water	x	x	
Marestail	x		
Marsh Yellowcress	x		
Pondweeds (various)	x		
Reed, Common	x	x	
Reed Grass	x		
Rushes (various)	x	x	
Sedges (various)	x	x	
Shoreweed	x		
Starwort, Common	x		
Stoneworts (various)	x		
Watercress	x	Water Crowfoot	x
Waterlilies			x
Water Milfoils (various)			
	x		
	x		

Appendix B

LATIN NAMES OF SPECIES MENTIONED

Mammals

Beaver, Canadian	*Castor canadensis*
Beaver, European	*C. fiber*
Capybara	*Hydrochoerus hydrochaeris*
Coypu	*Myocastor coypus*
Mink, American	*Mustela vison*
Muskrat	*Ondatra zibethicus*
Otter	*Lutra lutra*
Raccoon	*Procyon lotor*
Water Shrew	*Neomys fodiens*
Water Vole	*Arvicola terrestris*

Birds

Ducks:	
Carolina	*Aix sponsa*
Mallard	*Anas platyrhynchos*
Mandarin	*Aix galericulata*
Pochard	*Aythya ferina*
Shoveler	*Spatula clypeata*
Tufted	*Aythya fuligula*
Wigeon	*Anas penelope*
Canada Goose	*Branta canadensis*
Swan, Mute	*Cygnus olor*
Coot	*Fulica atra*
Cormorant	*Phalacrocorax carbo*
Dabchick	*Tachybaptus ruficollis*
Heron, Common	*Ardea cinerea*
Kingfisher, Common	*Alcedo atthis*
Moorhen	*Gallinula chloropus*
Reed Bunting	*Emberiza schoeniclus*
Sand Martin	*Riparia riparia*
Swallow	*Hirundo rustica*
Swift	*Apus apus*
Wagtail, Grey	*Motacilla cinerea*
Wagtail, Yellow	*M. flava*

Reptiles

Grass Snake	*Natrix natrix*
Terrapin, Caspian	*Clemmys caspica*
Terrapin, Reeves	*Geoclemys reevesii*
Terrapin, Spanish	*Clemmys leprosa*
Tortoise, American Pond	*Emy blandingii*
Tortoise, European Pond	*E. orbicularis*

Amphibians

Frog, Common	*Rana temporaria*
Newt, Common (smooth)	*Triturus vulgaris*
Newt, Great Crested	*T. cristatus*
Newt, Palmate	*T. helveticus*
Toad, Common	*Bufo bufo*
Toad, Natterjack	*B. calamita*

Fish

Bream, Common	*Abramis brama*
Bream, Silver	*Blicca bjoernka*
Carp, Common	*Cyprinus carpio*
Carp, Crucian	*Carassius carassius*
Carp, Grass	*Ctenopharyngodon idella*
Dace	*Leuciscus leuciscus*
Eel	*Anguilla anguilla*
Goldfish	*Carassius auratus*
Grayling	*Thymallus thymallus*
Gudgeon	*Gobio gobio*
Minnow	*Phoxinus phoxinus*
Orfe, Golden	*Leuciscus idus*
Perch	*Perca fluviatilis*
Pike	*Esox lucius*
Pike-Perch (Zander)	*Stizostedion lucio perca*
Roach	*Rutilus rutilus*
Rudd	*Scardinius erythrophthalmus*
Salmon	*Salmo salar*
Stickleback, 3-spined	*Gasterosteus aculeatus*
Tench	*Tinca tinca*
Trout, Brown	*Salmo trutta*
Trout, Rainbow	*S. gairdneri*

Bivalves

Cockle, Orb-shell	*Sphaerium* spp.
Cockle, Pea-shell	*Pisidium* spp.
Mussel, Painter's	*Unio pictorum*
Mussel, Pearl	*Margaritifera margaritifera*
Mussel, Swan	*Anodonta cygnaea*

Snails

Great Pond	*Lymnaea stagnalis*
Great Ram's Horn	*Planorbis corneus*
Ram's Horn	*P. planorbis*
Wandering	*Lymnaea peregra*

Crustaceans

Crayfish (Native)	*Austropotamobius pallipes*
Crayfish, 'Noble'	*Astacus astacus*
Crayfish, 'Signal'	*Pacifastacus leniusculus*
Freshwater Hoglouse	*Asellus* spp.
Freshwater Shrimp	*Gammarus* and *Crangonyx* spp.
Waterflea	*Daphnia* spp.
Water Slater	*Asellus aquaticus*

Other Invertebrates

Alderfly	*Sialis lutaris*
Backswimmer	*Notonecta glauca*
Bloodworm	*Chironomus plumosus* (larva of non-biting midge
Caddis Fly	Various species of order *Trichoptera*
Damselfly	Various species of suborder *Zygoptera*
Dragonfly	Various species of suborder *Anisoptera*
Great Diving Beetle	*Dytiscus marginalis*
Leech, Horse	*Haemopsis sanguisuga*
Leech	*Glossiphonia complanata*
Leech	*Erpobdella octoculata*
Leech, Medicinal	*Hirudo medicinalis*
Mayfly	Order *Ephemeroptera*
Pond Skater	*Gerris lacustris*
Rat-Tailed Maggot	*Eristalis* spp. (hoverfly larvae)
Sludgeworm	*Tubifex tubifex*
Water Boatman	*Coroxidae* spp.
Water Measurer	*Hydrometra stagnorum*
Water Scorpion	*Nepa cinerea*
Water Spider	*Argyroneta aquatica*
Whirligig Beetle	*Gyrinus natator*

Appendix C
ESTIMATING FLOW RATES OVER NOTCHED WEIRS

V-notch weir, right-angled

| Head | | Flow | |
Inches	Millimetres	Gallons/hour	Litres/hour
0.5	12.7	25	114
1.0	25.4	115	525
1.5	38.1	310	1,409
2.0	50.8	650	2,955
2.5	63.5	1,140	5,182
3.0	76.2	1,790	8,137
3.5	88.9	2,620	11,910
4.0	101.6	3,650	16,593
4.5	114.3	4,880	22,184
5.0	127.0	6,350	28,857
5.5	139.7	8,040	36,550
6.0	152.4	9,980	45,369
6.5	165.1	12,200	55,461
7.0	177.8	14,600	66,371
7.5	190.5	17,300	78,649
8.0	203.2	20,400	92,738
8.5	215.9	23,700	107,739
9.0	228.6	27,300	124,105
9.5	241.3	31,200	141,834
10.0	254.0	35,400	160,927
10.5	266.7	40,000	181,838
11.0	279.4	45,000	204,568
11.5	292.1	55,100	227,743
12.0	304.8	55,600	252,756

CONVERSION TABLE

1 inch = 25.4mm	1mm = 0.039in
1 foot = 30.48cm	1cm = 0.394in
1 yard = 0.914m	1m = 3.28ft/1.09yd

1 sq in = 6.452mm^2 $1m^2$ = 10.76sq ft
1 sq ft = 0.093m^2 $1m^2$ = 1.196 sq yd
1 sq yd = 0.836m^2 1 are = 100m^2
1 acre = 0.4047ha   1ha = 10,000m^2
     1ha = 2.471 acres

1 cu ft = 0.028m^3 $1m^3$ = 35.315 cu ft
1 cu ft = 28.32 litres $1m^3$ = 220 gallons
1 million gallons   1 litre = 0.22 gallons
   = 4,546 m^3 1,000 1 = 1m^3
   = 4.546 megalitres   1 million litres
     = 1 megalitre (Mg)

1 ounce = 28.35g   1 gramme = 0.035oz
1 pound = 0.454kg   1kg = 2.2046lb
1 cwt = 0.0508 tonnes (t)   1 tonne = 0.9842 tons
1 ton = 1.016t

Temperatures:

32°F = 0.0°C, 50°F = 10°C, 68°F = 20°C, 77°F = 25°C, 86°F = 30°C.

Approximate conversions

1 in / 25mm	1 litre per second / 13 gallons per minute
1 ft / 300mm	1m^3 per second / 13 gallons per minute
1m / 1.1yd	1m^3 per second / 2,000 cu ft per minute
1,000mm^2 / 1.5 sq in	30g / 1oz
1m^2 / 10 sq ft	1kg / 2lb
4ha / 10 acres	50kg / 1cwt
100cm^3 / 6 cu in	4.5kg per ha / 4lb per acre
3m^3 / 100 cu ft or 4 cu yd	5t per ha / 2 tons per acre
1m^3 / 220 gallons	11 litres per ha / 1 gallon per acre
9 litres / 2 gallons	

Appendix E
⚬⚬–ADDRESSES–⚬⚬

British Association for Shooting and Conservation
Marford Mill, Rossett, Wrexham, Clwyd LL12 0HL

British Crayfish Marketing Association Ltd
Riversdale Farm, Stour Provost, Gillingham, Dorset

British Herpetological Society
c/o Zoological Society of London, Regent's Park, London NW1 4RY

British Naturalists' Association
6 Chancery Place, The Green, Writtle, Essex CM1 3DY

British Trust for Conservation Volunteers
36 St Mary's Street, Wallingford, Oxon. OX10 0EU

British Trust for Ornithology
Beech Grove, Station Road, Tring, Herts. HP23 5NR

British Waterfowl Association
6 Caldicott Close, Over, Winsford, Cheshire CW7 1LW

Centre for Alternative Technology
Llwyngwern Quarry, Machynlleth, Powys SY20 9AZ

Council for the Protection of Rural England
4 Hobart Place, London SW7W 0HY

Countryside Commission
John Dower House, Crescent Place, Cheltenham, Glos. GL50 3RA

Department of the Environment
2 Marsham Street, London SW1P 3EB

Duncan & Associates
(manufacturers of oceanographic, limological and hydrobiological sampling apparatus)
Beech Court, Cart Lane, Grange-over-Sands, Cumbria LA11 7AF

Farming and Wildlife Advisory Group
The Lodge, Sandy, Bedfordshire SG19 2DL

Freshwater Biological Association
The Ferry House, Far Sawry, Ambleside, Cumbria LA22 0LP

The Game Conservancy
Burgate Manor, Fordingbridge, Hampshire

Green & Carter
(manufacturers of Vulcan rams)
Vulcan Iron Works, Ashbrittle, Wellington, Somerset TA21 0Q

Institute of Geological Sciences (Hydrogeology Unit)
Maclean Building, Crowmarsh Gifford, Wallingford, Oxon. OX10 8BB

Institute of Terrestrial Ecology
(including the Biological Records Centre)
Monks Wood Experimental Station, Abbots Ripton, Huntingdon, Cambridgeshire PE17 2LS

Institute of Water Engineers and Scientists
31–33 High Holborn, London WC1V 6AA

Intermediate Technology Publications Ltd
(publishers of *Appropriate Technology*)
9 King Street, London WC2E 8HN

Landscape Institute
12 Carlton House Terrace, London SW1Y 5AH

Local Council —
see local telephone directory

Ministry of Agriculture, Fisheries & Food —
see local telephone directory

National Farmers Union
Agriculture House, Knightsbridge, London SW1X 7NJ

National Federation of Anglers
Halliday House, 2 Wilson Street, Derby DE1 1PG

National Federation of Young Farmers Clubs
National Agricultural Centre, Kenilworth, Warwickshire CV8 2LG

National Trust
42 Queen Anne's Gate, London SW1H 9AS

The Otter Trust
Earsham, Bungay, Suffolk

Royal Society for Nature Conservation
22 The Green, Nettleham, Lincoln LN2 2NR

Royal Society for the Protection of Birds
The Lodge, Sandy, Bedfordshire SG19 2DL

Salmon and Trout Association
Fishmongers Hall, London Bridge, London EC4R 9EL

Soils and Water Management Association
National Agricultural Centre, Kenilworth, Warwickshire CV8 2LZ

Tiger Developments Ltd
(bridges, lakes and gazebos)
Deanland Road, Golden Cross, Hailsham, East Sussex BN27 3RP

Water Authorities Association
1 Queen Anne's Gate, London SW1H 9BT

Water Research Centre
Medmenham Laboratory, Henley Road, Medmenham, Marlow, Buckinghamshire SL7 2HD

Well Drillers Association
Orchard House, Foxton, Market Harborough, Leics. LE16 7RJ

Wildfowl Trust
Slimbridge, Glos. GL2 7BT

GLOSSARY

Acid: With a pH value of less than 7.0.

Aerobic: Requiring or having the presence of free oxygen.

Alkali: A substance which has more hydroxyl ions than hydrogen ions and has strong basic properties.

Alkaline: With a pH value of more than 7.0.

Anaerobic: Having or able to survive the absence of free oxygen.

Aquifer: 'Water-bearing', i.e. porous rock which holds and transmits water.

Artesian: A situation in which water is under pressure, confined beneath impermeable strata.

Base: A substance which reacts with an acid to form a salt and which dissolves in water to form hydroxyl ions.

Bog: Wet, peaty land with acid qualities.

Borehole: Small-diameter well sunk to great depth.

Calcareous: Relating to calcium carbonate (e.g. chalk and limestone); alkaline.

Carnivore: Creature which feeds on flesh.

Carr: Wet woodland or fen scrub, on which grows willow, alder, etc.

Catchment: Area from which rainwater drains to a watercourse, pond or wetlands.

Catch Pit: Chamber dug to collect spring water, etc.

Decomposer: Organism which breaks down dead plant and animal matter, releasing their constituent components.

Detritivore: Creature which eats detritus.

Detritus: Organic debris, i.e. decomposing plants and animals.

Dystrophic: Water of negligible productivity.

Eutrophic: Water of high productivity, well supplied with nutrients (literally, 'overfed').

Eutrophication: Process whereby the natural regression of a pond or lake to land is accelerated by overproduction.

Fen: Permanently wet peaty area, often in alkaline situation.

Groundwater: Water held in saturated rock.

Hardness: Quality of water containing dissolved calcium/magnesium salts (carbonates or sulphates).

Head: Measure of the potential energy of water created by the difference in depth at two given levels.

Herbivore: Organism that eats plant matter.

Impermeable: Not able to transmit water.

Ion: An electrically charged atom.

Marsh: Waterlogged soil, but not continuously flooded, and mineral based and often base-rich.

Mesotrophic: Of medium productivity.

Mire: Area of permanently wet peat.

Neutral Water: Water with a pH value of 7.0.

Oligotrophic: Of low productivity.

pH Value: 'Potential hydrogen', measured in terms of the reciprocal logarithm of the hydrogen ion concentration and expressing the relative acidity or alkalinity of a solution.

Peat: Soil composed entirely of organic remains (as opposed to minerals).

Perched Watertable: Saturated aquifer above the main watertable where the water is trapped by a small area of impermeable rock.

Permeable: Able to transmit water.

Photosynthesis: Process by which green plants (including algae) convert carbon dioxide and water into organic compounds (carbohydrates), using sunlight energy absorbed by chlorophyll. A byproduct of the process is oxygen.

Plankton: Floating microscopic aquatic plants and animals.
Phytoplankton consists of plants.
Zooplankton consists of animals.

Rhabdomancy: Water divination by means of a rod.

Riparian: Of or inhabiting a riverbank; owner of land bordering a river.

Run-off: Water flowing over the surface into a watercourse or pond.

Spillway: System or channel over which impounded water can flow to avoid flooding a dam.

Spring: Natural outflow of groundwater at the surface, at which point groundwater becomes surface water.

Surface Water: Water in watercourses, lakes and ponds.

Glossary

Swamp: Flooded or very wet land on mineral-based soils, usually under water in the growing season and dominated in most cases by emergent plants.

Toxic: Poisonous.

Watertable: Upper surface of waterlogged part of aquifer.

━━●●─BIBLIOGRAPHY─●●━━

Aslett, Ken, **Warwick**, John and **Boulders**, Jan: *Water Gardens* (RHS/Cassell, 1985)

Barrington, Rupert: *Making and Managing a Trout Lake* (Fishing News Books, 1983)

Belcher, Hilary and **Swale**, Erica: *A Beginner's Guide to Freshwater Algae* (HMSO/ITE, 1976)

Boyle, P.R.: *Molluscs and Man* (Edward Arnold, 1981)

Brassington, Rick: *Finding Water* (Pelham Books, 1983)

British Trust for Conservation Volunteers: *Waterways and Wetlands* (reprinted 1985)

Bryant, Paul, **Jauncey**, Kim and **Atack**, Tim: *Backyard Fish Farming* (Prism Press, 1980)

Chakroff, Marilyn: *Freshwater Fish Pond Culture and Management* (Vita Publications, 1976)

Clegg, John: *Freshwater Life* (Frederick Warner & Co., 1974)

Dipper, Dr Frances and **Powell**, Dr Anne (ed): *Field Guide to the Water Life of Great Britain* (The Reader's Digest Association, 1984)

Dyson, John: *Save the Village Pond* (Ford Motor Company, 1974)

Evans, John G.: *An Introduction to Environmental Archaeology* (Granada Publications, 1981)

Haslam, Sylvia, **Sinker**, Charles and **Wolseley**, Pat: *British Water Plants* (Field Studies Council 1982)

Hvass, Hans: *Reptiles and Amphibians* (Blandford Press, 1972)

Johnson, A.A. and **Payn**, W.H.: *Ornamental Waterfowl* (Saiga, 1979)

Kabisch, Klaus and **Hemmerling**, Joachim: *Ponds and Pools* (Croom Helm, 1984)

Keswick, Maggie: *The Chinese Garden* (Academy Editions, 1978)

Macan, T.T. and **Worthington**, E.B.: *Life in Lakes and Rivers* (Collins, 1974)

Maitland, P.S.: *Hamlyn Guide to Freshwater Fishes of Britain* (Hamlyn, 1977)

Mason, I.L. (ed.): *Evolution of Domesticated Animals* (Longman, 1984)

Mellanby, H.: *Animal Life in Freshwater* (Chapman & Hall, 1975)

Muhlberg, Helmut: *Complete Guide to Water Plants* (EP Publishing, 1982)

Philips, Roger and **Rix**, Martyn: *Freshwater Fish of Britain and Europe* (Pan Books, 1985)

Quigley, Michael: *Invertebrates of Streams and Rivers: A Key to Identification* (Arnold, 1977)

Spencer-Jones, D. and **Wade**, M.: *Aquatic Plants: A Guide to Recognition* (ICI, 1986)

Swift, Donald R.: *Aquaculture Training Manual* (Fishing News Books, 1985)

Swindell, Philip: *Water Gardening* (Michael Joseph, 1975)

Thompson, Gerald, **Coldrey**, Jennifer and **Bernard**, George: *The Pond* (Oxford Scientific Films/William Collins, Sons & Co., 1984)

Vince, John: *Discovering Watermills* (Shire Publications, 1980)

Wilson, Ron: *The Back Garden Wildlife Sanctuary Book* (Astragal Books, 1979)

Young, Geoffrey: *The Sunday Times Countryside Companion* (Country Life Books, 1985)

Also, essential booklets issued by the following organisations:

Farming & Wildlife Groups (farm conservation ponds)

The Game Conservancy (management of wildfowl waters)

Ministry of Agriculture, Fisheries & Food (technical publications relating to irrigation, dam building, drainage systems, watercress culture, farm pond management, etc.)

Nature Conservancy Council (conservation of amphibians, dragonflies, molluscs and other species; river plant communities, etc. — send for full list of publications)

Water Authorities (regional)

INDEX

Index